Cross Country Pipeline Risk Assessments and Mitigation Strategies

Cross Country Pipeline Risk Assessments and Mitigation Strategies

Arafat Aloqaily, PhD
International expert in process safety and risk management

Gulf Professional Publishing is an imprint of Elsevier
50 Hampshire Street, 5th Floor, Cambridge, MA 02139, United States
The Boulevard, Langford Lane, Kidlington, Oxford, OX5 1GB, United Kingdom

The author's royalties will be donated in total to charities supporting cancer patients in need. Purchasing your copy of the book will directly contribute to this cause.

Library of Congress Cataloging-in-Publication Data
A catalog record for this book is available from the Library of Congress

British Library Cataloguing-in-Publication Data
A catalogue record for this book is available from the British Library

ISBN: 978-0-12-816007-7

For information on all Gulf Professional publications
visit our website at https://www.elsevier.com/books-and-journals

Publisher: Susan Dennis
Acquisition Editor: Anita Koch
Editorial Project Manager: Carly Demetre
Production Project Manager: Maria Bernard
Cover Designer: Greg Harris

Typeset by SPi Global, India

Dedication

"To my parents for their love and support"

Contents

Disclaimer

This book is presented solely as a general guideline for pipeline risk assessments. The information provided is not meant to be used, nor should it be used in lieu of applicable regulations, good engineering practices, and standards. While best efforts have been used in preparing this book, the author and publisher make no representations or warranties of any kind, assume no liabilities of any kind with respect to the accuracy or completeness of the contents, and specifically disclaim any implied warranties of use for a particular purpose. Neither the author nor the publisher shall be held liable or responsible to any person or entity with respect to any loss, incidental or consequential damages caused or alleged to have been caused, directly or indirectly, by the information or programs contained herein. Every pipeline situation is different, and the advice and strategies contained herein may not be suitable for every situation. Readers should seek the advice of a competent professional before implementing any improvement program based on this book. References are provided for informational purposes only and do not constitute endorsement of any websites or other sources. Readers should be aware that the content of the websites listed in this book may change. Efforts have been made to reference material used from other sources. However, if the reader recognize a material source or copyright that has not been appropriately referenced, please contact the author or publisher to include it in the next revision.

PipeFAIT

PipeFAIT is an innovative tool designed to assess pipeline failure modes and probabilities based on pipeline specific design and operating conditions. The tool is easy to implement and robust enough to assess potential integrity issues for pipelines. Overall, the tool has more than one million combinations of different possibilities and can provide recommendations to mitigate pipeline risk and failure causes. It can be accessed at

https://www.elsevier.com/books-and-journals/book-companion/9780128160077

Chapter 1

Foreword and Book Description

Pipeline risk assessments are very essential to understanding the actual magnitude of risks posed by pipelines carrying hazardous materials to the general public and environment and pipeline operating personnel. This book is mainly intended as reference that comprehensively describes risk assessment approaches and methodologies in a simple and direct format. The scope of this book covers the cross-country pipelines that are defined as pipelines that transfer hazardous material such as oil and gas products from the production source to the processing/distribution sites. Cross-country pipelines are pipelines that start just outside the limits (fence line) of the production site all the way to the entry/fence line of the receiving end. Pipes or pipelines within the fence line of production or processing facilities are not included in this definition of cross-country pipelines and are not within the scope of this book. Cross-country pipelines are not used to mean international pipelines only, although pipelines that run across international borders are considered cross-country pipelines. This definition is very important to understand while going through the book. Also, it is important to note that offshore pipelines are beyond the scope of this book, although some information and principles discussed in this book are applicable to those as well.

Pipelines are widely considered a safe mode for transporting hazardous material, and the pipeline network has been consistently growing in size throughout the world in the last few decades. Maintaining this vast network of important assets presents a serious challenge to owners of these assets and all other stakeholders involved including the local communities and authorities. The main issue that consistently manifests itself is the lack of credible data that allow for proper evaluation of potential release of hazardous material from these pipelines and assessing the risk from the pipelines. Loss of containment (LOC) of materials transferred via the pipelines, as a result of pipeline failure, is the main event that should be controlled to maintain the safety and protect the environment around communities where these pipelines run. Pipeline failure modes and integrity assessment are critical components in maintaining pipeline safety and managing its risk. This cannot be more critical than when these pipelines carry flammable/toxic material commonly processed/produced from chemical, petrochemical, oil, and gas facilities. In this book, the potential modes and causes of failure of pipelines used in the oil and gas industry are evaluated based on wide range of data available from different databases in the world that cover more than 40 years of operating history. The objective of the analysis is to develop a consistent approach that allows for

Cross Country Pipeline Risk Assessments and Mitigation Strategies
https://doi.org/10.1016/B978-0-12-816007-7.00001-9

proper estimation of potential risk and how it can be mitigated. This will then be combined with consequence modeling to fully calculate the different forms of risk presented by the pipelines. Simplified consequence modeling for pipeline risk assessments is presented in this book, as well.

Pipeline professionals and experts must understand the pipeline safety and risk, just the way they understand pipeline design and operations. Some references are available that talk about this concept, but this book describes the fundamentals of pipeline risk management in simple straightforward manner. It provides a simplified model that allows the readers to use the model and get a reasonably detailed assessment of the risk. It also describes general principles on risk control and mitigation. The book presents a detailed description of pipeline failure modes and a simple, yet accurate, approach to evaluate that based on publicly available information. The intent of the book is not to describe theories behind risk assessment but to present a simple easy-to-follow model that allows the reader to understand and assess risk associated with the pipelines.

BACKGROUND AND HISTORIC PERSPECTIVE

Pipelines are used to transfer hazardous material in large quantities between sources of supplies and consumers/end users throughout the world. Pipelines are effective and economic mean of transfer for hazardous material and are gaining momentum as the preferred transporting tool. Figs. 1–4 show the growth in pipeline network total length in Europe and the United States with time. The data shown in these figures demonstrate the growing reliance on pipelines to transfer hazardous material between different points of interest.

Fig. 1 shows that the total inventory of pipelines transferring hazardous liquid material in the United States increased from 270,000 km in 2004 to around 320,000 km in 2014 while the total gas pipeline length increased from 3.3 million km in 1995 to less than 4 million km in 2014 [1]. The European liquid pipeline inventory increased from 14,000 km in 1970 to 34,000 km in 2014, as

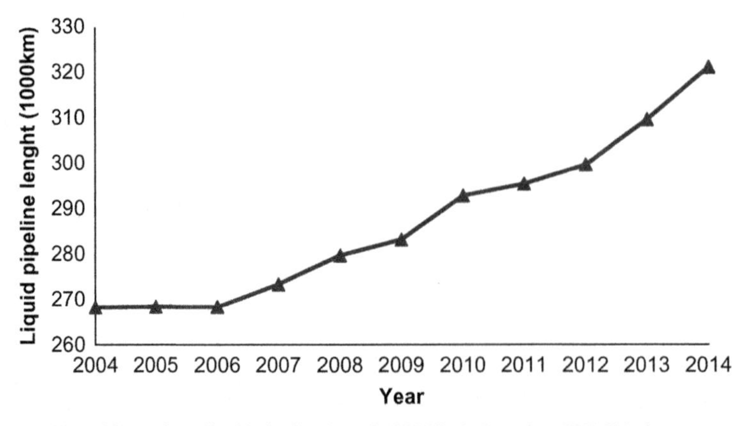

FIG. 1 US total hazardous liquid pipeline length (1000 km). Based on PHMSA data.

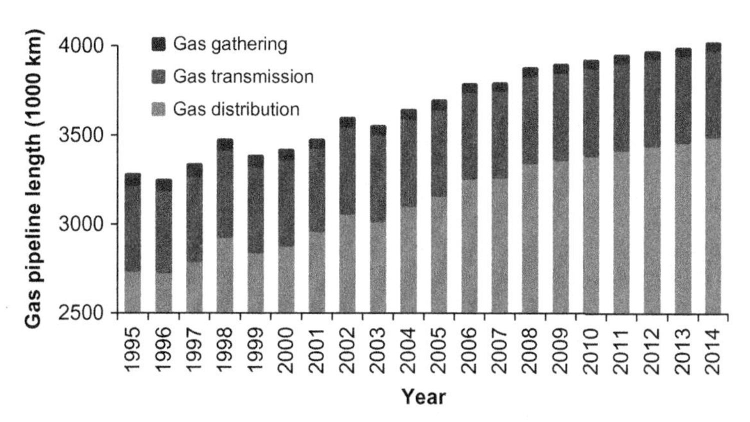

FIG. 2 US total gas pipeline length (1000 km). Based on PHMSA data.

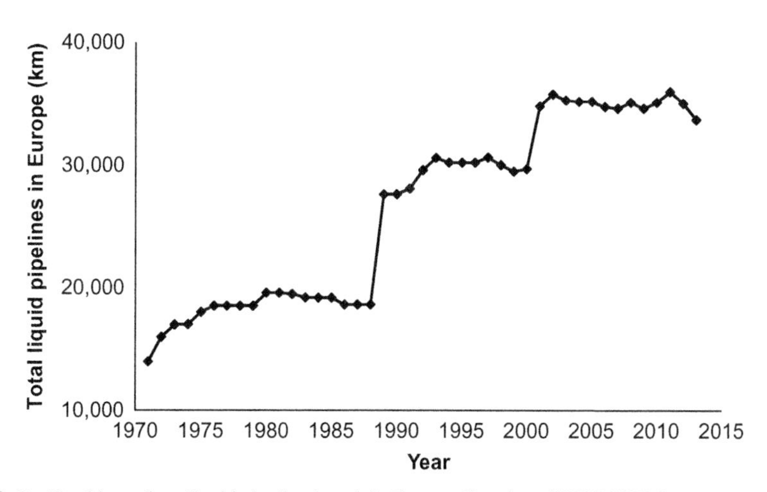

FIG. 3 Total hazardous liquid pipeline length in Europe. Based on CONCAWE data.

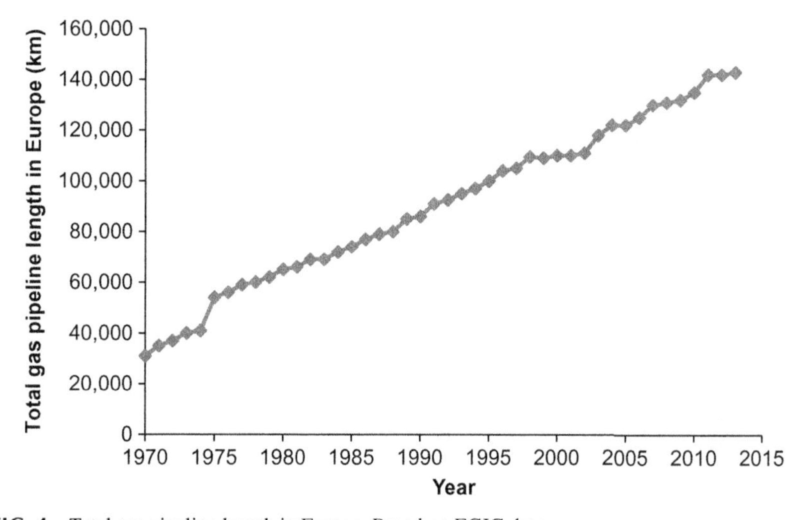

FIG. 4 Total gas pipeline length in Europe. Based on EGIG data.

shown in Fig. 3, while the total gas pipeline length increased from 30,000km in 1970 to 140,000km in 2013 [2,3].

The trend shown in Figs. 1–4 indicates that more pipelines will be constructed and used to transport hazardous material around the world. The large network of pipelines poses risk to public communities and the environment where these pipelines pass. The risk should be assessed and managed to acceptable levels in order to maintain proper balance between public safety and environment protection in one side and economic development from the other side.

PIPELINES DESIGN AND OPERATION

Pipelines are designed and operated/maintained for one purpose, which is to transfer hazardous material from one location (source or production facility) to another location (end user or consumer stations/facilities). In order to achieve this objective, the pipeline has to be designed and operated per applicable standards and best practices to ensure that its design is appropriate and its operation is conducted to keep it running per design conditions.

Pipeline design includes selecting/calculating the following parameters:

Pipeline Size (Diameter)

This parameter is determined mainly by the amount of material to be transferred through the pipeline (i.e., flowrate). The higher the flowrate is, the bigger the pipeline size is. Industrial best practices provide proper guidelines on acceptable flowrates for a given pipeline cross-sectional area (i.e., velocities) for both gases and liquid fluids. These recommended velocities are a balance between the size of the pipeline and its operational need. If too low velocities are used, then larger pipelines will be needed, which increases the cost of construction and causes operational problems (e.g., accumulation of liquid in low points leading to corrosion problems). High velocities reduce the size of the pipeline but could cause other problems such as erosion, damage to pipeline material, and high pressure loss inside the pipeline. So, it is recommended to follow the applicable standards and best practices to ensure optimum design of the pipeline and proper selection of its size.

Recommended ranges of optimum velocities for different fluids are available in the literature and could be used for pipeline design as applicable.

Pipeline Wall Thickness

The thickness of the pipeline walls depends on several factors including the following:

- The operating pressure and temperature: The higher the pressure and temperature is, the higher the required wall thickness should be. High pressure requires higher thickness to ensure that the pipeline does not rupture.

- Corrosivity of the material and the required corrosion allowance: Highly corrosive material requires thicker pipeline to ensure that corrosion does not reduce the thickness to the point where a leak occurs. Note that corrosion allowance is not always used in cross-country pipelines.
- Design factor: This is a factor used to increase the wall thickness if the pipeline carries hazardous material that can impact the public. The wall thickness increases to ensure safety of the public. Increase in the wall thickness is proportional to the size of public communities exposed to hazardous materials in case of pipeline rupture. Note that design factors can also be modified by changing the operating pressure of the pipeline as well.
- Material of construction can also affect the pipeline wall thickness as well. Stronger material (higher grade) can reduce the wall thickness, but the ratio between the diameter and wall thickness has a limit that it should not exceed.

Wall thickness for pipelines is calculated following international best practices and standards such as the ASME 31.4 and ASME 31.8 [4,5].

Pipeline Material Grade

The selection of pipeline material grade depends on the operating and design factors including the pressure, temperature, and required corrosion resistance. International and industrial best practices provide detailed specification for the selection of pipeline material. API, ASME, ASTM, and ANSI standards can be used for this purpose. Economy is a deciding factor as well.

Pipeline External Coating

External coating is used to protect pipelines from external corrosions especially when the pipeline is buried, where external corrosion can be an issue. Different types of coatings have different abilities to reduce external pipeline corrosion. The right coating type depends on many factors including the characteristics of the backfill (i.e., nature of the soil surrounding the buried pipeline) and the overall corrosion management approach such as the use of cathodic protection. Certain coatings are more effective than others.

Determining Whether Pipeline is to be Buried or Not

Pipelines are typically buried to protect them from external impact and damage. However, sometimes, burying a pipeline or a segment of it is not an option. For example, when the land or soil is highly corrosive, it might be better not to bury the pipeline as external corrosion of the pipeline might be a problem. Also, when pipelines are close to valve stations or pump/compressor stations (used to boost the pipeline pressure), then pipelines will be placed aboveground. If the pipeline is expected to require a lot of maintenance and regular work, then burying it will be impractical. The actual decision to bury a pipeline or not shall be made on a case-by-case basis and depend on the design intent and operating conditions. Note that the overwhelming majority of the cross-country pipelines are buried.

Pipeline Route and Design Factor

The route of the pipeline should be selected to optimize the economy and risk/safety of the pipeline. The route should be chosen to be as short as possible to reduce capital cost invested in constructing the pipeline. However, sometimes, the shortest pipeline route may not be the best choice from risk/safety perspective as it could be increasing the risk to the public or the environment. If the pipeline route passes by large public communities, then the design of the pipeline shall be adjusted to mitigate the risk posed by the pipeline. One of the most effective ways to do so is to adjust the design factor of the pipeline to increase its wall thickness (among many other measures that will be discussed later in this book). There are different pipeline classes associated with different population densities around the pipeline route. Increasing the pipeline class through reducing the pressure inside the pipeline or increasing its wall thickness can improve the safety and mitigate the risk by reducing the pipeline vulnerability to corrosion and external damage. These classes are associated with different design factors. Details on determining the design factor and pipeline class can be found in open literature.

PIPELINES SYSTEMS

Cross-country pipelines are comprehensive engineering systems designed to transport hazardous material in a safe manner from source point to destination. Typically, the system comprises the following elements:

Main Pipeline

This refers to the actual pressurized vessel (pipe) that transfers the hazardous material from one location to the other. Typically, this is the main component of the pipeline system. It should be designed, constructed, operated, and maintained/inspected according to applicable codes and standards/practices to ensure that the pipeline operation is smooth and the risk posed by the pipeline is managed appropriately. This element of the pipeline is the most expensive, and due to the fact that the pipeline passes through vast area, it is the segment that requires a lot of attention to keep the pipeline safe.

Metering Stations

In order to determine the amount of fluid flowing in the pipeline, a metering station is used, which has a flowmeter of some kind suitable for the fluid being pumped in the pipeline. Typically, metering stations are placed at least at the source and sometimes at the destination as well. This allows the pipeline operator and its customers to know how much fluid is being transferred and to perform a material balance for the purpose of identifying if a leak occurs in the pipeline. Metering stations can cause some pressure loss and have to be accounted for in the design.

Details on metering station design and operations can be found in the literature such as *Pipeline & Gas Journal* [6].

Scraping Launcher and Receiver

In order to maintain the integrity of the pipeline and prevent its failure, the pipeline must be inspected and maintained appropriately. One of the essential techniques to achieve is scraping the pipeline. Different types of scraping operations are used, the most common of which are the following:

- Intelligent scraping used to measure the corrosion rate and whether it compromised the pipeline thickness
- Regular scrapping used to clean the pipeline

The scraper is a piece of equipment that is inserted into the pipeline, through a launching pad called scraper launcher, when it flows through the pipeline all the way to the end, where it is collected in the scraper receiver station. The scraper flows along with the fluid and cleans the pipeline or collects information on pipeline wall thickness, which is used to evaluate the need to fix pipeline segments that are losing thickness due to corrosion.

Pipeline scraping is an essential part of the pipeline operation and is conducted at a predetermined frequency to ensure the integrity of the pipeline. Typically, cleaning scraping is run much more often than intelligent scraping (which also cost more than cleaning scraping). Useful information is available in open literature on scraping operations and scraper design. Scraping is also referred to as pigging in many literature and sources, and the name comes from the shape of the equipment itself.

While scraping is a simple concept in theory, its operation can be risky and complicated. It involves opening the scraper launcher that is connected to pressurized pipelines and loading the scraper. Then closing the launcher and sending the scraper to the receiver end where the receiver is opened and the scraper is retrieved. The launcher and receiver should be isolated before they are opened, and there are safety mechanisms and procedures to ensure safe operation. However, taking shortcuts when conducting the operation or the presence of faulty equipment that gives wrong measurements of pressure can cause serious safety incidents that can lead and has led to fatalities. Scraping operations shall be conducted by well-trained and adequately supervised individuals following well-written/clear procedures to reduce the likelihood of these incidents.

Good information on pipeline scraping operation can be found in open literature [7].

Pressure-Boosting Stations

For the fluid inside the pipeline to flow from one location (starting point or source) to other end (destination), a pressure gradient has to exist. The source has to have pressure high enough to push the fluid to the destination. While the fluid flows inside the pipeline, pressure loss will occur due to friction between the fluid and pipe walls as well as other pressure loss elements such as valves, fittings, and elbows; the pressure at the source should be at or above the pressure at the destination plus the pressure loss inside the pipeline. For long pipelines, increasing the pressure at the source to meet this requirement might not be feasible (technically and economically) due to the increased wall thickness required for the pipeline and size of pressure-boosting equipment at the source. The alternative would be to use pressure-boosting stations along the pipeline to boost pressure when it drops below critical value. This way, the fluid can be transferred for long distances without excessive pressure increase at the source. Pressure-boosting stations can be pump stations for liquids or compressor stations for gases.

Determining when to use pressure-boosting stations, how many is needed, and where to locate them is a complicated decision that depends on many factors including the economy of pipeline operations versus capital cost of the pipeline and the station equipment, as well the required pressure profile along the pipeline rate.

Leak Detection Systems

Pipeline leak detection systems are engineering systems used to detect leak of materials from the pipeline, in order to alert the operator to leak incidents. Leak detection is an essential component of pipeline risk management as it allows the operator to respond in time to the leaks to prevent further escalation of incidents. Different technologies are available to detect the leak from pipelines, depending on the nature of the fluid in the pipeline and the leak size. These range from basic material balance techniques to much more complicated systems. A good description of different technologies and applications for pipeline leak detection systems is available in the literature [8–10].

Although pipeline leak detection systems can be quite sophisticated and take a lot of effort to operate and maintain, they may not always be very effective. Some assessments in the United States found that leak detection system effectiveness is less than 20% only [11]. The published data in Europe's pipeline incident reports show that the majority of leaks are determined by methods other than the engineering leak detection system [2,3]. However, leak detection systems can still be useful in picking up some leaks and may still be worth implementation from risk management perspective. The higher the risk posed by the pipeline, the more sophisticated and important leak detection systems should be.

Storage Facilities and Receiving Stations

In order to ensure smooth operations and no interruption to material supplies and for flexibility of operation and maintenance activities, pipeline operators maintain storage facilities at the delivery location or other locations along the pipeline route as needed. The storage facilities must contain storage tanks and some pumping station. The capacity is determined by the customer needs and operator philosophy of operation.

INTEGRITY MANAGEMENT SYSTEMS

Maintaining the pipeline integrity is essential to managing risk and maintaining supply continuity. Preventing pipeline failure is the main objective of integrity management activities/systems. Two essential systems are discussed below:

Corrosion Management Systems

Since pipelines are constructed from metals (mainly steel in most cases), they are vulnerable to corrosion from the surrounded environment (especially if the pipeline is buried) and from the material carried inside the pipeline. Corrosion can reduce the wall thickness and cause leaks/failures, which increase the risk of the pipeline and compromise its integrity.

Corrosion management is an essential part of pipeline operation. Corrosion is managed typically through several measures including the following:

- Selection of proper material of construction that is compatible with the surrounding environment and fluid being transferred to make the pipeline material suitable for corrosion resistance.
- Using proper coating material to reduce external corrosion, especially for the buried pipeline, where external corrosion could be an issue.
- Applying cathodic protection that is used to reduce corrosion by applying electric current to the pipeline to counter the corrosion driving force and protect the pipeline.
- Injecting corrosion inhibitor inside the pipeline to control corrosion due to the material being transferred in the pipeline. Corrosion inhibitor is injected typically at the source at a rate sufficient to control the corrosion in the pipeline.
- Managing the change in fluid characteristics to ensure that it does not deviate from design limits to the point that it becomes incompatible with pipeline material from corrosion perspective.

Surge Managing Systems

To protect the pipeline against sudden changes in pressure that could damage the pipeline and affects its integrity, surge protection systems should be used.

Surge (i.e., sudden increase in the hydraulic pressure in the pipeline that could exceed its maximum allowable pressure) could lead to pipeline failure. It is mitigated through several measures including the use of special relief valves called surge relief valves. Regular pressure relief mechanisms (such as regular relief valves) could also be used and can be effective against surge.

Inspection

To maintain the integrity of the pipeline and assess the effectiveness of measures being used, inspection must be conducted on a regular basis. This includes performing activities to measure the corrosion rates and assess the mechanical integrity of the pipeline as a whole. One of the most common techniques used is the scraping activities known as in-line inspection (ILI), which measures the wall thickness of the pipeline and evaluates its corrosion rate. Also, other inspection techniques can be used for the other component of the pipeline system mentioned earlier in this chapter.

Preventative Maintenance

Preventative maintenance (PM) refers to the activities conducted to maintain pipeline components and systems following a preset program in order to uncover weak points and fix them before they lead to failure. This could include fixing the points where the wall thickness is getting thin due to corrosion before they leak, replacing a damaged valve, or eroding mechanical seal on a pump. PM is essential to ensuring and maintaining a good integrity of the pipeline system.

Management of Change (MOC)

Any change introduced to the pipeline after it is designed and operated should be assessed and must not introduce new hazards or risks. This is done through a proper management of change (MOC) process that evaluates the hazards introduced with the change and how they are mitigated. For example, increasing operating pressure or flowrate in the pipeline can expose the pipeline to higher stress. If this is not evaluated and mitigated appropriately, it can compromise the pipeline integrity and eventually lead to failures increasing the risk posed by the pipeline. Proper MOC practices are an essential component of a robust pipeline integrity and safety management programs.

PIPELINES LIFE-CYCLE MANAGEMENT

Proper risk management means that the pipeline safety and integrity should consider a life-cycle management approach. This means that safety and risk mitigation should be part of the entire life cycle of the pipeline from design all the way to decommissioning, including construction, maintenance, and operation. Throughout these different phases, all activities and aspects should be conducted according to acceptable industry best practices and governing regulations. The following are examples of how to achieve that:

- Design stage: proper design is essential to maintain safety and reduce risk of the pipeline. The following points should be observed:
  - Adequate sizing of the pipeline will help reduce erosion, corrosion, and potential pressure surge. An oversized pipeline will increase corrosion, and undersized pipeline will increase erosion. Both compromise the integrity and can damage the pipeline leading to failure and hazardous incidents.
  - Proper wall thickness of the pipeline will protect the pipeline against external impact from activities near the pipeline and against corrosion as well.
  - Proper selection of the pipeline material will help control corrosion. The use of wrong material can expedite corrosion and lead to pipeline failure.
- Construction/commissioning stage: incorrect construction activities such as installation, welding activities, and incorrect application of coatings can cause damage to the pipeline that will compromise its integrity and cause operational problems and lead to hazardous events/incidents. Also, choosing the route that reduces exposure to public facilities/communities and sensitive environment will reduce the risk posed by the pipeline.
- Operation and maintenance: this phase constitutes the main part of the pipeline life cycle. A lot of attention is paid to this stage for controlling the pipeline risk. Following proper operations, maintenance, and inspection practices will ensure adequate risk management and pipeline integrity.
- Decommissioning stage: once the pipeline is no longer needed, the proper decommissioning activities will ensure that the hazard is removed and the pipeline no longer poses any risk.

GEOPOLITICAL PERSPECTIVE

Pipelines that are constructed to transfer hazardous material across international boundaries represent a unique situation, where the geopolitical environment could be a significant factor in determining the mode and continuity of operation. Most of these pipelines are constructed to transfer oil and gas from producing countries to consuming markets/countries. Typically, the pipelines are constructed based on international (bilateral or multilateral) agreements between the countries where these pipelines run and operate. The impact of geopolitics on these agreements and arrangements can be high. Political tension could also affect operation continuity as has been seen in several cases in recent years.

There are several pipelines connecting Asia with Europe or running through North America, for example. In all cases, constructing/running international pipelines not only can enhance cooperation between nations but also can bring some tension between others. It is beyond the scope of this book to discuss this in detail, but the reader can refer to the literature for material published on this topic.

REFERENCES

[1] USA-DOT, Pipeline and Hazardous Materials Safety Administration data on pipeline incidents. https://www.phmsa.dot.gov/.
[2] Concawe Oil Pipelines Management Group, Performance of European Cross-Country Oil Pipelines Statistical Summary of Reported Spillages in 2014 and Since 1971, 2016.
[3] EGIG Gas Pipeline Incidents, 9th Report of the European Gas Pipeline Incident Data Group (period 1970–2013), Doc. Number EGIG 14.R.0403, February, 2015.
[4] NSI/ASME Standard B31.4, Liquid Transportation Systems for Hydrocarbons, Liquid Petroleum Gas, Anhydrous Ammonia, and Alcohols.
[5] ANSI/ASME Standard B31.8, Gas Transmission and Distribution Piping Systems.
[6] S. Mokhatab, Fundamentals of Gas Pipeline Metering Stations, vol. 236, no. 1. Tehran Raymand Consulting Engineers/Greg Lamberson International Construction Consulting, LLC, Iran/Tulsa, OK USA, 2009.
[7] J. Cordell, H. Vanzant, Pipeline Pigging Handbook, third ed., Clarion Technical Publishers, 2003.
[8] M. Henrie, P. Carpenter, R. Edward Nicholas, Pipeline Leak Detection Handbook, first ed., Gulf Professional Publishing, 2016.
[9] K.E. Abhulimen, A.A. Susu, Liquid pipeline leak detection system: model development and numerical simulation, Chem. Eng. J. 97 (1) (2004) 47–67.
[10] P.-S. Murvay, I. Silea, A survey on gas leak detection and localization techniques, J. Loss Prev. Process Ind. 25 (6) (2012) 966–973.
[11] PHMSA, Report on Leak Detection, NY Time Report, 22nd December, 2012.

Chapter 2

Identification of Hazards Associated With Pipelines

Pipelines carrying hazardous material can pose hazard and risk to surrounding communities, the environment, public facilities, and industrial assets. For the purpose of this book, the main focus will be on flammable and toxic hazardous material such as the material transferred in the oil and gas industry and petrochemical and chemical industries. Example of the material includes the following:

- Upstream oil and gas material
  - Crude oil produced from the field and transferred from producing sites to processing facilities. This can be sour or sweet but could carry water and sand with it. It could also contain associated gas as well.
  - Nonassociated gas produced from upstream producing sites and transferred to processing facilities. This can be sour (toxic) or sweet (not toxic).
- Stabilized crude oil from processing facilities where gas has been mostly removed and the gas content is typically low compared with unstabilized crude.
- Refined petroleum products such as gasoline, kerosene, diesel, or heavy gas oil.
- Sweet sales gas or natural gas (NG), which is constituted of methane mainly with traces of heavier materials such as ethane and propane.
- Natural gas liquid (NGL).
- Other hazardous materials such as ammonia, chlorine, and CO_2.

Sour pipelines contain H_2S or carbon dioxide. H_2S in particular is a highly toxic material that can travel far from the source in case of release and cause serious injuries and possible death to exposed personnel. These materials pose flammable and toxic hazard, which can manifest itself in the following format:

- Flammable hazards
  - Fire: pool fire, jet fire, and flash fire
  - Thermal radiation associated with fires
  - Blast (or explosion) hazard
- Toxic hazards from the exposure to the toxic material such as ammonia, hydrogen sulfide, and carbon dioxide.[1]

1. CO_2 is more of an asphyxiating agent not a toxic one.

https://doi.org/10.1016/B978-0-12-816007-7.00002-0

Hazards are defined as the inherent capacity to cause harm. In other words, if something can cause harm by nature, then it is hazardous regardless of whether this hazard materializes or not. Examples of hazards include electricity, transportation, flammable/thermal, and toxic hazards. For example, a toxic material is hazardous, whether someone gets exposed to it or not. Same thing applies to electricity and other hazards. The harm/damage can only occur after being exposed to the hazard.

Hazard identification (HAZID) and evaluation is a key component in quantifying and managing risk of pipelines. In order to effectively deal with hazards, they must be identified in a systematic manner following well-established procedures and practices. Generally, in this process, a hazard is recognized, its causes are identified, and its potential consequences are evaluated and documented. This is a process that starts in early stages of the project and continues throughout the life cycle of the system.

Typically, a workshop is conducted by a multidisciplinary team that includes representatives from all stakeholders including inspection, maintenance, engineering, process and operations, and other relevant groups to perform the HAZID exercise. Risk reduction and control measures will then be determined and prioritized by the team.

Hazards to be identified include anything that can potentially cause harm or any possible events and conditions that could lead to hazardous situations for personnel, the environment, or assets. This also includes operational impacts, business interruption, downtime, repairs, and inventory loss. Each identified hazard should be given an ID number. At the same time, causes, consequences, and existing safeguards for each hazard should be noted. Following that, a qualitative risk assessment is conducted for the identified hazards [1].

Depending on the type of hazard available and the purpose of the assessment, different HAZID techniques can be used. Common techniques used in the industry are described in this chapter.

PRELIMINARY HAZARD ANALYSIS

Preliminary hazard analysis (PHA) is an initial high-level screening exercise that can be used to identify, describe, and rank major hazards during conceptual stage of a facility design. This technique can also be used to identify possible consequences and likelihood of occurrence and provide recommendations for hazard mitigation following the general HAZID approach described above.

PHA does not relatively take long time to be conducted and provides information early in the life cycle of the project. However, it remains as a simple technique that is not systematic, depends on the team's experience, and is used mainly where limited information is available about the pipeline design [2]. As seen in Table 1, this assessment identifies the type of hazard associated with pipeline (i.e., leak of hazardous material) and defines the potential consequences and mitigation measures. It also provides recommendations and crude qualitative assessment of risk based on the risk matrix (as described in Chapter 1). Likelihood (L) and severity (S) values used in Table 1 here reflect conditional modifiers in them (such as ignition probabilities).

HAZARDS AND OPERABILITY ANALYSIS

This is a HAZID technique that focuses mainly on the pipeline components and aims at enhancing/improving its design to run safely and reducing the inherent risk of the pipeline (that is coming from pipeline design). The hazard and operability (HAZOP) technique is a structured procedure and attempts to identify how a process may deviate from the design intent. The emphasis in the HAZOP study is on identifying potential problems, not necessarily solving them. If, however, a solution is obvious during the study, it is recorded or incorporated immediately while the study is in progress. This technique depends on the use of a "guide word" related to the design intent, then identifying how the process can deviate from that.

In HAZOP studies, a multidisciplinary team systematically "brainstorms" the process under review in a series of meetings using a set of guide words to structure the review. The team is composed of individuals representing a variety of specialties (as described in the section above). This multidisciplinary team concept allows the various viewpoints of the team members to stimulate the thinking of the other team members and results in creative thinking. Consequently, a more thorough review is achieved than would occur if members of the team individually reviewed the same process.

For the HAZOP assessment to be useful and comprehensive, the following conditions must be satisfied:

- Obtain a full description of the process, including the intended design conditions.
- Systematically examine every part of the process to find out how deviations from those designed or specified intent can occur.
- Decide whether these deviations can give rise to hazards and/or operability problems.

Details on HAZOP methodology are available in published guidelines and literature [3–6]. Table 2 shows example of some guide words used in HAZOP. More guide words can be used as needed. Fig. 1 illustrates the following general steps during HAZOP studies, based on the methodology described in the referenced guidelines [3].

For pipelines, the combination of parameters and guide words given in Table 3 can be used. Table 4 shows an example of a liquid pipeline HAZOP worksheet. The items shown in the sheet are for illustration purposes and are not conclusive. The actual HAZOP worksheet shall be developed using the actual P&ID drawings of the pipeline following the HAZOP methodology. However, these examples shown in Table 4 still provide an idea about pipeline HAZOP worksheets.

TABLE 1 Example of PHA Assessment for Pipeline Carrying Hazardous Material

Type of hazard	Causes	Consequences	L[a]	S[b]	R[c]
Leak of hazardous material to atmosphere	Pipeline overpressurization (e.g., closing isolation valve by mistake, scraper stuck in pipeline, and pressure control system malfunction)	Pool fire	M[d]	L[e]	L
		Flash fire		M	H[i]
		Jet fire		M	M
		Explosion		H	M
		Toxic		H	H
	Corrosion	Pool fire	L	L	L
		Flash fire		M	L
		Jet fire		M	L
		Explosion		H	M
		Toxic		H	M
	External impact	Pool fire	H	L	M
		Flash fire		M	M
		Jet fire		M	M
		Explosion		H	H
		Toxic		H	H
	Operational errors (e.g., hot tapping by error)	Pool fire	M	L	L
		Flash fire		M	H
		Jet fire		M	M
		Explosion		H	M
		Toxic		H	H
		Pool fire	L	L	L
		Flash fire		M	L
		Jet fire		M	L
		Explosion		H	M
		Toxic		H	M

[a] *Likelihood (low, medium, or high).*
[b] *Severity (low, medium, or high).*
[c] *Risk (low, medium, or high).*
[d] *Medium level.*
[e] *Low level.*
[f] *Emergency shutdown systems.*
[g] *High-integrity pressure protection systems.*
[h] *Leak detection systems.*
[i] *High level.*

Mitigation	Recommendations	Action Owner
• Pump ESD[f] system • Overpressure protection systems • Operating procedures (closing and opening isolation valves and scraping procedures) • Pressure alarms and indicators on pipeline • Control ignition sources • Use PPE (fire resistance clothes—FRC—and Scott airbags for toxic impact) • Emergency response planning (ERP) • Spill control procedures	• Consider installing redundant pressure protection systems (e.g., HIPPS[g] or relief valves) • Ignition sources control around pipeline • Design drainage system around pipeline to prevent the formation of large pools • Install LDS[h] for early detection of leak • Install toxic/flammable detectors around pipeline, where it passes by close proximity of public communities to initiate ERP	Engineering department
• Corrosion management program (e.g., corrosion inhibitors, external coating, and cathodic protection systems) • Scarping and in-line inspection (ILI) • Proper material grade selection • External inspection of pipeline	• Maintain design velocity to avoid water accumulation in low points of pipeline (that can cause corrosion) • Consider increasing corrosion allowance in pipeline material • Eva)luate using corrosion resistance alloy (CRA) material to control corrosion	Operations, inspection, and engineering departments
• Buried pipeline (if aboveground) • External protection using concrete slabs (for buried segments of pipeline) and crash barriers (for aboveground pipeline segments)	• Develop proper excavation procedures and work permit process to control third-party activities near the pipeline • Establish designated pipeline corridor and protect it through fences and patrolling	Operations and security departments
Operating procedures	Establish and enforce hot work permit	Operations department
Security procedures and patrol of pipeline corridor	Consider installing CCTV in areas close to public communities where impact can be high for continuous monitoring of pipeline (especially for aboveground segments)	Security department

TABLE 2 Example of Guide Words Used in HAZOP

Guide Word	Meaning	Comments
No	Complete negation, for example, of intention	The lack of any relevant physical property such as "no forward flow" when there should be flow
More	Quantitative increase	More of any relevant physical property than there should be (e.g., higher flow, temperature, pressure, viscosity, heat, and reaction)
Less	Quantitative decrease	Less of any relevant physical property than there should be (e.g., lower flow, temperature, pressure, viscosity, heat, and reaction)
As well as	Quantitative increase	All design and operating intentions are achieved together with some additional activities (e.g., impurities and extra phase)
Part of	Quantitative decrease	Only some intentions are achieved; some are not
Reverse	Opposite of intention	Reverse flow or chemical reaction (e.g., inject acid instead of alkali in pH control)
Other than	Complete substitution or miscellaneous	No part of the original intention is achieved; something quite different occurs. Also start-up, shutdown, alternative mode of operation, catalyst change, corrosion, etc.

WHAT IF ANALYSIS

What if analysis (WFA) is a HAZID technique based on answering questions derived from the team's imagination concerning failure scenarios and their consequences. This method is generally used for simple facilities such as product storage facilities. A version of this technique is called what-if/checklist, where a checklist is used to ask questions instead of freestyle brainstorming exercise.

This is a flexible, quick, and easy-to-perform technique that can be conducted at any stage of the pipeline's life cycle and does not require advanced set of skills to be conducted. However, it is not systematic to some degree, and the outcome depends mainly on the experience of the team members participating in the study.

A what-if form, consisting of columns assigned to identify the item under consideration, lists the question, describes the potential consequence/hazard, and lists the recommendations. Additionally, columns can be employed to assign work and to indicate completion.

FIG. 1 Illustration of HAZOP procedure.

TABLE 3 Guide Words for Line-by-Line Analysis for Pipelines

Guide Word				
Flow	High	Low	Zero	Reverse
Pressure	High	Low		
Temperature	High	Low		
Impurities	Gaseous	Liquid	Solid	
Two-phase flow				

TABLE 4 Example of Liquid Pipeline HAZOP

Deviation	Cause	Consequences	Safeguards	S[a]	L[b]	R[c]	Recommendation	Action Owner
No flow	Pump failure	Production losses	Standby spare pump with automatic start-up	H	M	M	Not needed	
	Pipeline rupture	Leak of hazardous material into the atmosphere causing harm to humans and the environment and business losses	• ESD • ERP	H	L	M	Install LDS for early detection of leaks to minimize consequences	Engineering department
	Isolation valves on the pipeline closed by mistake	Loss of production and potential overpressurization of pipeline	• Pump ESD system • Pressure gauges on pipeline and pump	H	L	M	• Install valve position monitoring systems to continually monitor the position of the valve and get early notification about valve position • Lock open the valve and use operating procedures to ensure it only closes when needed	Engineering and operating departments
Less flow	Pipeline leak	Loss of production and potential exposure of personnel and the environment to hazardous material	• ESD • Flow measurement element • ERP	L	M	L	• Install LDS for early detection of leaks to minimize consequences • Consider setting ESD to shutdown at low pressure signals	Engineering and operation departments
	Pressure/flow controller malfunction	Loss of production and impact on operation of pipeline	• Redundant pressure/ flow indicators	M	L	L	• Consider using more pressure and flow indicators as needed	Engineering department

Reverse flow	Pipeline rupture at upstream point Back pressure from downstream point	Loss of production/ operability issue	• Check valves to prevent reverse flow	H	L	M	None	
More flow	Pressure/flow controller malfunction	Increased pressure in pipeline with potential to cause failure of the pipeline More flow can cause operability issues	• Flow and pressure indicators on pipeline downstream of pump • Proper maintenance and inspection of control systems • ESD	M	M	M	Evaluate increasing storage capacity downstream of pipeline to absorb extra flow	Engineering department
No pressure	Failure of pump	Production losses	Standby spare pump with automatic start-up	H	M	M	Not needed	
	Pipeline rupture	Leak of hazardous material into the atmosphere causing harm to humans and the environment and business losses	• ESD • ERP	H	L	M	• Install LDS for early detection of leaks to minimize consequences • Consider setting ESD to shutdown at low pressure signals	Engineering and operation departments
Less pressure	Pressure controller malfunction	Loss of production and impact on operation of pipeline	Redundant pressure/flow indicators	M	M	M	• Consider using more pressure and flow indicators as needed • Consider setting ESD to shutdown at low pressure signals	Engineering department
	Pipeline leak at upstream point	Loss of production and potential exposure of personnel and the environment to hazardous material	• ESD • Flow measurement element • ERP	H	L	M	• Install LDS for early detection of leaks to minimize consequences	Engineering department

(*Continued*)

TABLE 4 Example of Liquid Pipeline HAZOP—cont'd

Deviation	Cause	Consequences	Safeguards	S[a]	L[b]	R[c]	Recommendation	Action Owner
More pressure	Pressure controller malfunction	Potential impact on the operation of pipeline if pressure increases beyond MAOP[d] of the pipeline	• Redundant pressure/ flow indicators • Maintenance and inspection programs of pipeline control systems • ESD	M	L	L	• Consider using more pressure and flow indicators as needed • Evaluate using a fully rated pipeline to handle maximum possible pressure (if feasible) • Install pressure relief system on pipeline	Engineering department
	Blockage downstream of pipeline (e.g., scraper getting stuck in pipeline and isolation valves closed by mistake)		• Scraping procedures • Operating procedures for isolation valves • Locked open isolation valves (if needed) • ESD	M	L	L	• Consider installing valve position indicators to monitor valve position • Install pressure indicators on pipeline to monitor pressure and avoid increase in pressure beyond MAOP • Evaluate using a fully rated pipeline to handle maximum possible pressure (if feasible)	Engineering department
More temp.	Change in atmospheric temperature	Potential cause of two-phase flow inside the pipeline that could increase vibration of pipeline and potential damage to its support causing a failure of the pipeline	• Temperature indicators • Pipeline operating and inspection procedures	L	M	L	Insulating pipeline to prevent temperature increase Consider using surge drum to remove gases and maintain single phase flow (if needed)	Engineering and inspection department
Less temp.	Hydrate formation causing blockage in pipeline (less flow and high pressure)	See consequences for less flow and more pressure above		M	M	M	Insulate pipeline Consider use of hydrate formation inhibitors	Engineering department

[a] *Severity (low, medium, or high).*
[b] *Likelihood (low, medium, or high).*
[c] *Risk (low, medium, or high).*
[d] *Maximum allowable operating pressure.*

Table 5 shows a simple form of the what-if table that can be implemented in the analysis. A more formalized checklist analysis can also be conducted. Questions can be formulated around human errors, process upsets, and equipment failures during normal operations, maintenance activities, and other situations [7].

For pipelines, the following questions are examples of what can be asked[2]:

- What if the pipeline leaks
- What if high-pressure flammable, corrosive, or toxic gas leaks into a liquid pipeline
- What if the pipeline is fractured
- What if the pipeline becomes brittle
- What if the pipeline support fails
- What if the pipeline get impacted by internal or external forces
- What if the pipeline is subject to backflow
- What if the pipeline is subject to flow or pressure surge
- What if the pipeline is subject to liquid hammer
- What if the pipeline is subject to vibration
- What if fittings on the pipeline (e.g., gasket, seals, or flanges) leak
- What if pressure relief not provided or fail
- What if hydrate forms inside the pipeline
- What if pipeline pressure increases beyond maximum allowable operating pressure (MAOP)
- What if scraper get stuck inside pipeline
- What if corrosion protection systems fail
- What if emergency isolation mechanisms fail
- What if leak detection systems malfunction
- What if excavation procedures not followed
- What if preventative maintenance not performed

BOW-TIE ANALYSIS

Bow tie is a HAZID technique that helps build a visual relationship between the causes of an event on one side and its consequences on the other side. It also shows the prevention and mitigation measures for dealing with the causes and consequences of the events, respectively. These are called barriers as they are thought to prevent top event (leak of hazardous material from the pipeline in this case) from happening and prevent escalation of top events once they happen.

Bow-tie analysis is an adaptation of three conventional system safety techniques: fault tree analysis (FTA), event tree analysis (ETA), and causal factor charting [2]. It is very useful for establishing the cause-and-effect relationship and focuses on the main event assessment. Fig. 2 is an example of bow-tie analysis for pipeline leak (loss of containment, LOC) event.

2. These questions are from Ref. [2].

TABLE 5 What-If Analysis Form for Pipelines

What If?	Answer	L[a]	S[b]	Recommendations
Pipeline leaks	• Potential exposure of personnel and to hazardous material (e.g., causing fire or toxic impact) • Potential escalation of events (such as ignition of flammable releases causing fire or explosion)	Medium	High	• Use pipeline leak detection systems (LDS) to detect leaks and isolate pipeline early • Reduce potential development around pipeline to minimize potential exposure to the pipeline hazards • Install emergency isolation valves (EIVs) on the pipeline to effectively isolate leaks and reduce potential size of leaks • Prepare emergency response plan (ERP) for all cases where there is potential leak of hazardous material into the atmosphere
Pipeline becomes brittle	• Weakening of pipeline material strength with potential rupture of pipeline	Low	High	• Ensure that material selection of the pipeline is done to avoid embrittlement • Follow other measures indicated for pipeline leaks as necessary/feasible • Follow measures indicated for pipeline hydrate formation below as necessary
Pipeline support fails	• Pipeline might fracture or get damaged	Low	High	• Ensure that design of pipeline support is adequate for the pipeline load and potential vibration • Consider routine inspection and maintenance of pipeline support to ensure its integrity • Implement measures to prevent external impact on pipeline support by operators or third-party groups
Pipeline get impacted by external forces	• Pipeline can be damaged requiring repair	Low	Medium	• Consider protecting the pipeline from external impact • If pipeline is buried use concrete slabs with clear markers to prevent damage during excavation • If pipeline is aboveground, use crash barriers to prevent impact on the pipeline from vehicles • Consider burying aboveground pipelines
	• Pipeline can leak	Low	High	

Pipeline subject to backflow	• Loss of production • Back pressure on upstream facilities	Low	Low	• Use check valves to prevent backflow and flow indicators to measure flow and provide information if flow is going in the right direction • If needed, consider double check valves to reduce likelihood of backflow even further
Pipeline is subject to flow or pressure surge	• Can cause pipeline damage	High	Medium	• Install and maintain pipeline surge protection systems • Ensure that pipeline design can handle such events
Pipeline subject to vibration	• Can cause damage to pipeline and its support • Potential leaks of hazardous material in severe cases of vibration	Medium	High	• Install vibration control systems along the pipeline • Monitor pipeline vibration levels and rectify any issues immediately
Hydrate forms inside the pipeline	• Potential blockage of pipeline • Restriction of flow • Overpressure of pipeline upstream of hydrate formation location causing rupture and leak of pipeline material	Low	High	• Use hydrate formation inhibitor • Consider heating or insolating pipeline where feasible (if source of hydration is external) • Consider using pipe-in-pipe with heating fluid in between the two pipes (especially for offshore subsea pipelines)
Pressure increases beyond MAOP	• Potential rupture of pipeline causing leak of hazardous materials	High	High	• Implement pressure control systems to limit the possibility of overpressurizing the pipeline • Use overpressure protection systems such as emergency shutdown systems (ESDs) and high-integrity pressure protection systems (HIPPS) • Consider using fully rated pipelines to avoid overpressurization scenarios if feasible

(Continued)

TABLE 5 What-If Analysis Form for Pipelines—cont'd

What If?	Answer	L[a]	S[b]	Recommendations
Scraper get stuck inside the pipeline	• The loss of production • Potential overpressurization of the pipeline	High	Medium	• Use scraper passage indicators to monitor scraper location • Increase flow to push scraper if possible but without causing damage to the pipeline • In extreme cases (if pressure increases too much), then consider cutting pipeline and removing the scraper
Pressure relief system fails	• Potential overpressurization of pipeline causing damage to pipeline and leak of hazardous material	Medium	High	• Implement pressure control systems to limit the possibility of overpressurizing the pipeline • Use pressure indicators to monitor the pressure along the pipeline and avoid overpressurization scenarios • Use overpressure protection systems indicated above • Prepare ERP as needed

[a] *L is likelihood.*
[b] *S is severity.*

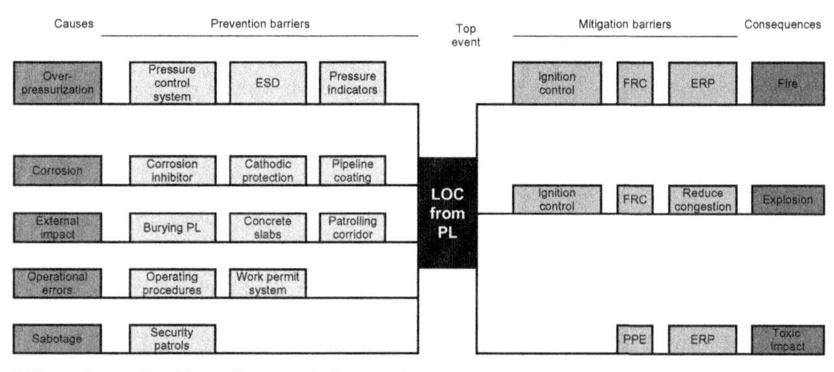

FIG. 2 Example of bow-tie analysis for pipeline.

Bow tie can also be used as a semiquantitative risk assessment tool. Frequencies of initiating events (causes) and probabilities of preventative and mitigating barriers would be used in combination with the bow-tie approach to estimate the potential damage caused by the different outcomes of the top event. Combining that with the consequences of the different outcomes leads to the risk associated with each outcome, and the summation of that can yield the total risk associated with the top event.

FAULT TREE ANALYSIS

FTA is a method used normally for evaluating detailed component failure. This involves using a logic diagram or tree to establish the various sequences of events that are required to reach some ultimate event (top event). With individual component failure rate data, this technique can be used to quantify the probability of a given event (top event) or subevents (main events that lead to the top event).

FTA is a top-down failure analysis technique, where the undesired state of a system is analyzed using Boolean logic to combine a series of lower-level events [8]. This method is applicable to understand how systems can fail, and hence engineer/design risk reduction measures, or to determine the failure rate of a particular system level/component.

FTA can be used to

- understand the sequence of events causing the top event (i.e., undesired state) and prioritize the factors leading to it,
- establish compliance with systems safety and reliability requirements and monitor/control the safety performance of the complex system,
- serve as a design tool that identifies engineering and reliability requirements to maintain the safety of the system being analyzed, and
- function as a diagnostic tool to identify and correct the causes of the top event. It can help with the creation of diagnostic manuals.

Basic diagram of FTA is illustrated in Fig. 3. Applying this approach to pipeline failures is demonstrated in Figs. 4–6. As shown, the concept is very helpful to identify all potential causes and paths leading to the top event, and it could be applied to a wide range of cases.

The example shown in these figures is for demonstration purposes and is not necessarily comprehensive. Only detailed FTA was developed for corrosion and overpressurization primary events. Similar approach can be used for other three events shown in Fig. 4. In Fig. 6, only high-integrity pressure protection system (HIPPS) failure was analyzed in detail for demonstration purposes, but failure of other layers of protection can be further analyzed as well. Several references are available on FTA and recent development in this technique [2,4,9].

FTA can be used to estimate the total probability of the top event occurring and show the contribution of each main/primary event. The following is an illustration of how the probability is calculated for the ESD failure (Fig. 7).

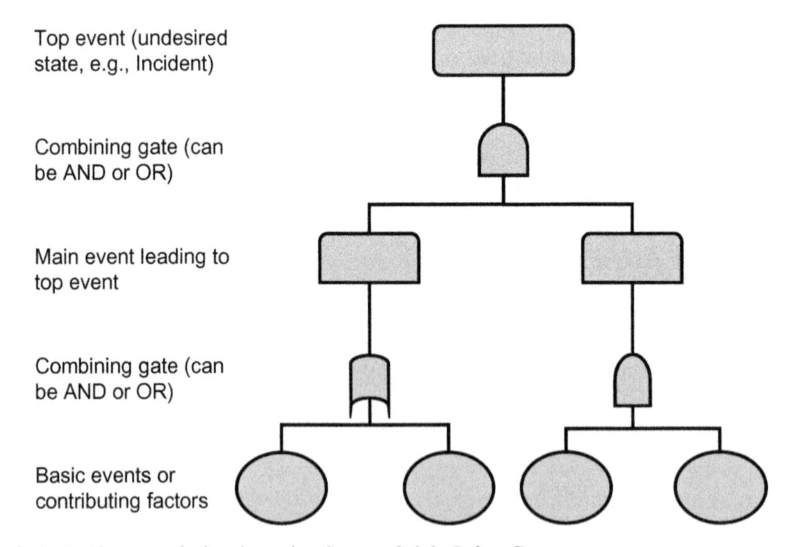

FIG. 3 Fault tree analysis schematic. *Source*: Geigle Safety Group.

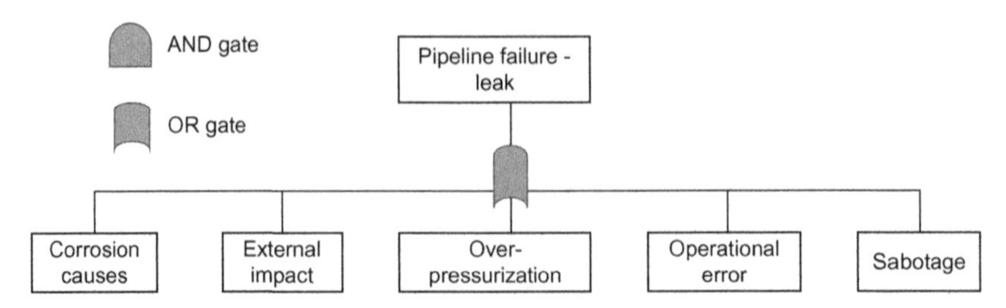

FIG. 4 Fault tree analysis example for pipeline failure.

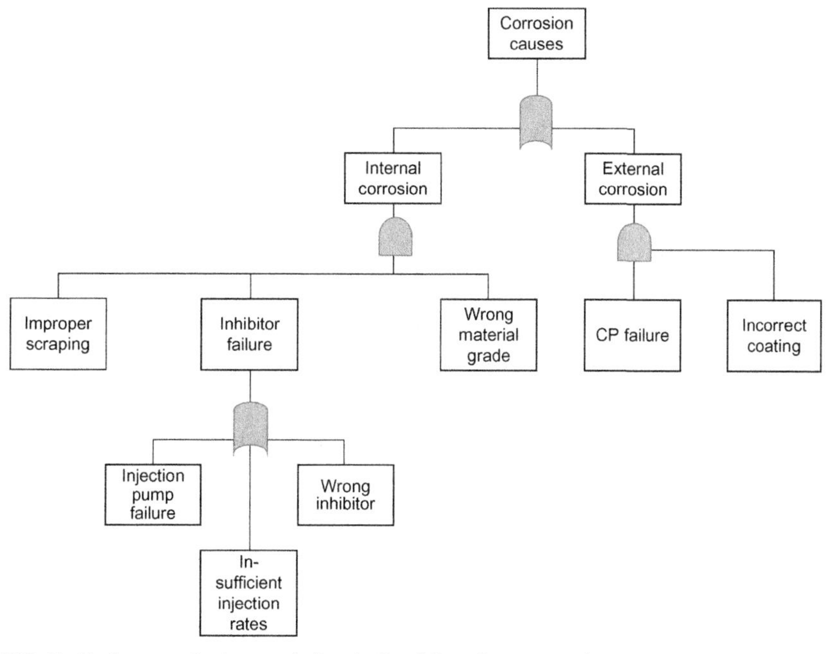

FIG. 5 Fault tree analysis example for pipeline failure due to corrosion.

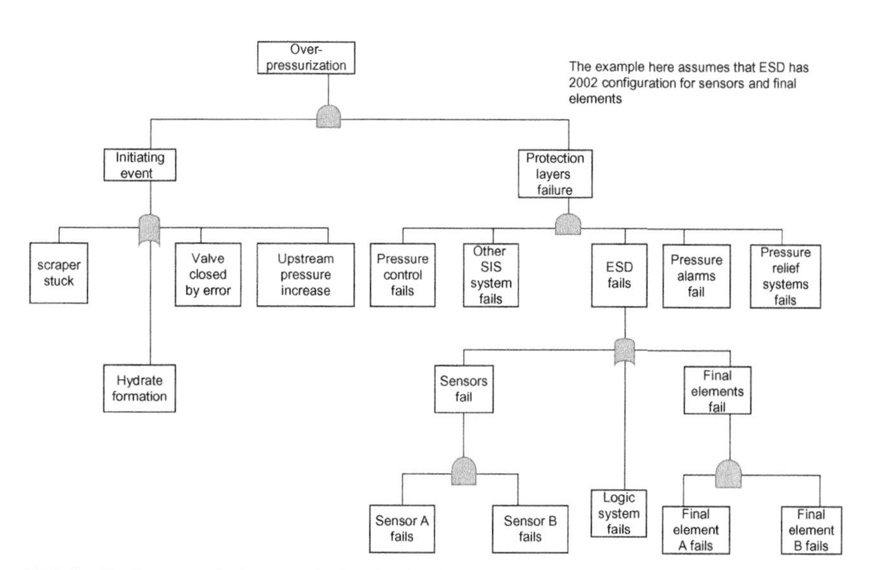

FIG. 6 Fault tree analysis example for pipeline failure due to overpressurization.

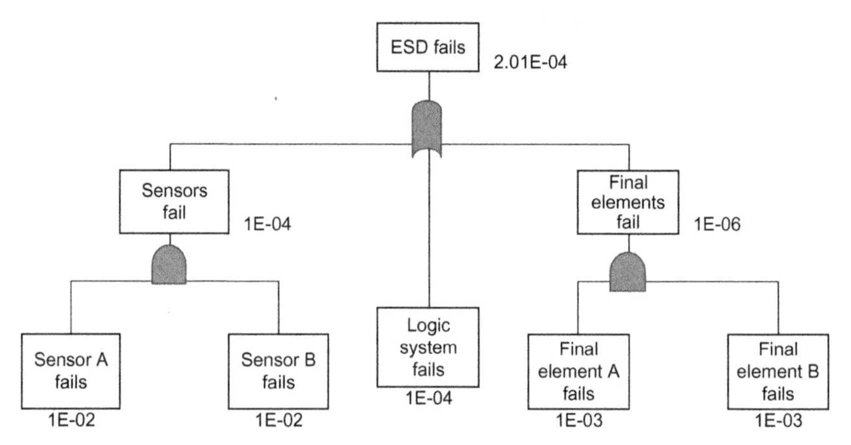

FIG. 7 Example of failure probability assessment using FTA for ESD system.

The numbers in the chart above represent the probability of component/system failure in a year. Note that the example shown here is a simplified one and is mainly used for illustration purposes. More detailed analysis can be performed depending on the purpose and availability of information.

EVENT TREE ANALYSIS

ETA is a method for providing the logical consequences of an event (such as leaks from pipelines). This frequently results in diagrams that can aid in reviewing interrelationships between events and how a primary event leads/escalate into another.

In ETA, a series of subsequent events leading to given outcomes are analyzed. Unlike FTA, ETA is used to evaluate all potential outcomes of an event, rather than starting with the primary event and assessing all subevents. ETA does generate the following information [10]:

- A qualitative relationship between top/primary event and subsequent outcomes resulting from that event provided that certain conditions are met (also called conditional modifiers).
- Using the qualitative description generated in the ETA, values can be assigned to different conditional modifiers and initial/primary events to calculate the possibility/likelihood of outcome scenario (subevent) taking place.
- Just like FTA, where dominant subevent can be evaluated quantitatively, ETA can define the dominant outcome scenario in terms of likelihood of occurrence.
- For pipeline risk, ETA can provide a quantitative and qualitative description of all potential hazards and how they escalate to different scenarios. This can then be used for pipeline risk assessments and mitigation.

Fig. 8 summarizes the steps for developing an ETA in general. A general ETA for pipeline events (i.e., leaks) is summarized in Fig. 9, and an example of pipeline ETA is given in Fig. 10.

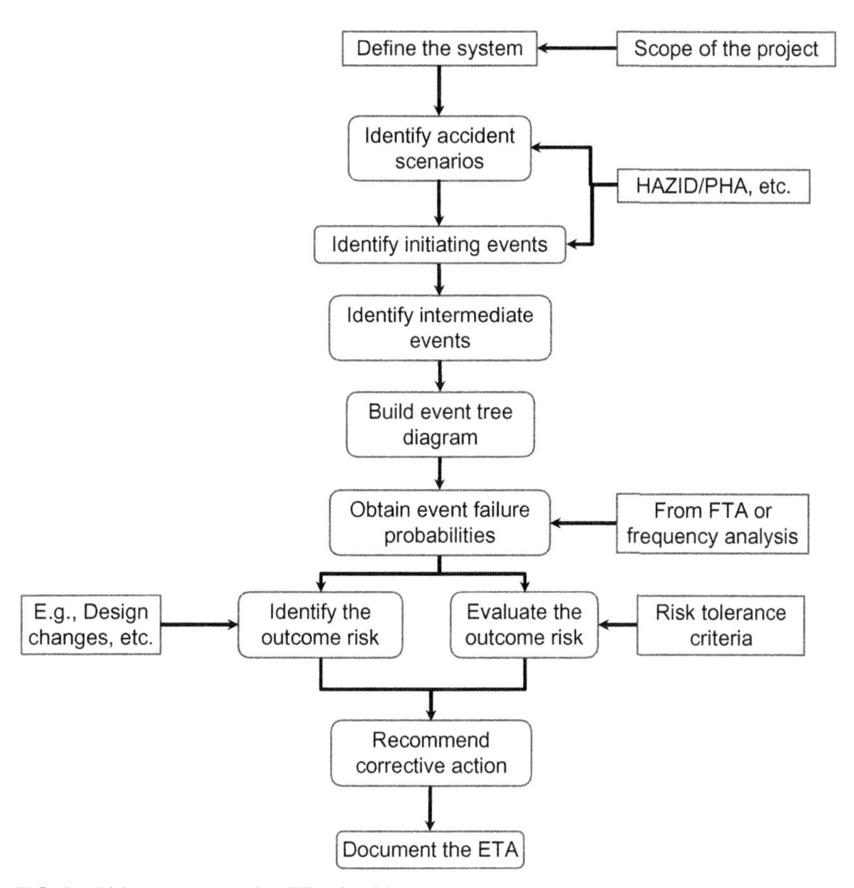

FIG. 8 Risk assessment using ETA algorithm.

Initiating event	Material phase	Contains toxics?	Early ignition?	Late ignition?	Congestion & confinement?	Consequence outcome
Loss of containment from pipelines	Gas	No	No	No		Gas release (pollution)
				Yes	No	Flash fire
					Yes	Explosion
			Yes			Jet fire
		Yes	No	No		Toxic exposure
				Yes	No	Flash fire
					Yes	Explosion
			Yes			Jet fire
	Liquid	No	No	No		Oil spill (pollution)
				Yes	No	Flash fire/pool fire
					Yes	Explosion
			Yes			Jet fire/pool fire
		Yes	No	No		Toxic exposure
				Yes	No	Flash fire
					Yes	Explosion
			Yes			Jet fire/pool fire

FIG. 9 ETA block diagram.

Top event	Leak size	Early ignition	Delayed ignition	Congestion/ confinement	Outcome scenario	Probability of all events combined	
Pipeline leak 1.0E-04	0.5 inch 0.5	Yes 0.01			Jet fire 5.00E-07		
		No 0.99	Yes 0.1	No 0.9	Flash fire 4.50E-06	5.0E-05	
				Yes 0.1	Explosion 5.00E-07		
			No 0.89		Toxic impact 4.45E-05		1.0E-04
	2.0 inch 0.25			Same as above		2.5E-05	
	6.0 inch 0.15			Same as above		1.5E-05	
	FBR 0.1			Same as above		1.0E-05	

FIG. 10 Event tree for pipelines.

The numbers used in Fig. 10 are for illustration purposes only. Detailed ignition probabilities and leak frequencies are discussed in Chapter 5. In the example shown here, the initiating event of a pipeline leak is 1.0E−4 leak/km/year, which is distributed between different release sizes as follows:

- 0.5 in release: 50% of total releases
- 2.0 in release: 25% of total releases
- 6.0 in release: 15% of total releases
- Full bore rupture (FBR) that is large enough to be equivalent to the pipeline size: 10% of total releases

Conditional modifiers of the ignition probabilities and presence of congestion/confinement are presented as shown in the figure. The final frequency of the outcome scenario is the multiplication of all initiative frequencies and the probabilities of all conditional modifiers in the path leading to the outcome scenario. For example, for jet fire, the pipeline leak frequency of 1E−4 is multiplied by the 0.5 in release probability and the early ignition probability of 0.5 and 0.01, respectively. The jet fire frequency is then estimated as 5E−7 event/pipeline km/year. The same applies for the rest of the events. The total frequency of all scenarios must add up to the initiating event, which is shown in the figure.

The relationship between the FTA and ETA is in the initiating event frequency of the ETA, which can be generated by FTA. As shown in Fig. 4, all events/causes leading to pipeline failure can be assessed, and the probability of leak is evaluated using FTA technique. Then, this can be used in the ETA as an initiating event frequency to determine the frequency of outcome scenarios.

So, these two techniques complement each other, just like all other HAZID techniques. Detailed information on this technique is also presented in other Refs. [11,12].

For a bow-tie analogy, the fault tree would be on the left side of the bow tie, and the event tree would be on the right side of the bow tie. FTA is used to evaluate all causes and their probabilities, and ETA is used to assess the consequences of the top event and their likelihood given all conditional modifiers and enabling conditions.

OTHER HAZID EXAMPLES

As indicated earlier, the pipeline systems include components other than the main pipes, such as storage tanks and pumps, scraper launchers, and receivers. Below are some examples of HAZID assessments for some of these components. It is important to keep in mind that these are used just to illustrate the basic concepts. Detailed assessment should be done based on the design and available information of the system being evaluated.

Pipeline Pump Example

Fig. 11 is an illustration diagram of a pump used to transfer the fluid through the pipeline. The source of the fluid is the storage tank typically available in the upstream facility connected to the pipeline. A process control system is available to ensure that the pump operates within design limit, and a protection

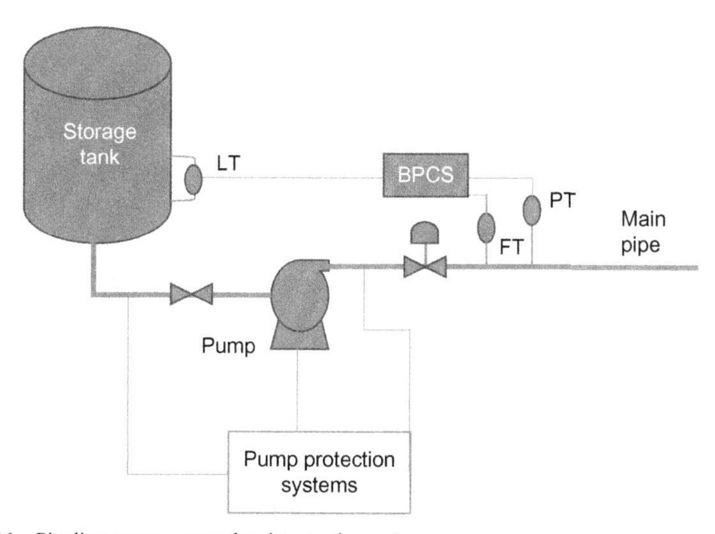

FIG. 11 Pipeline pump—control and protection system.

system is used to protect the pump from damage should the operating parameters exceed the design limits. Typically, the following control and protection layers are available:

- Control systems, which are intended to control the pump operations within design limit. They are part of the regular operation of the pump and are not intended to initiate pump shutdown for protection/safety purposes:
  - Level control to prevent the pump from running dry that causes pump cavitation leading to pump damage
  - Pressure control to ensure that the discharge pressure does not exceed acceptable limits leading to pump damage or pipeline failure
  - Flow control to ensure that the flow does not exceed design limit
- Protection systems, which will initiate pump shutdown to prevent damage to the pump or the pipeline:
  - Very low-level control to protect against pump cavitation
  - Very high-pressure protection system
  - High-temperature protection system to protect against damage when the pump runs dry
  - Vibration protection system to protect against mechanical damage of the pump resulting from excessive vibration of the pump due to mechanical or operational causes

When conducting HAZID assessment of the pipeline, the entire systems is divided into nodes. The pump here is used as one node. An example of HAZOP assessment report for the pump node is given in Table 6. Other questions can be found in literature for pumps [2]. Fault tree analysis example for temperature increase in the pipeline pump is shown in Fig. 12. More details can be added once the actual P&ID information is available. This example is only given for illustration purposes. FTA for other issues, such as pump damage due to pressure increase, can be assessed following the same approach.

Pipeline Scraper Launcher Example

Fig. 13 is an illustration diagram of scraper launcher used to launch the scraper through the pipeline. In principle, this is a simple operation, yet can be hazardous. The following steps summarize the procedure:

- The scraper launcher door is opened after the scraper is depressurized and drained safely.
- Scraper is loaded to the launcher, and the door is closed.
- The launcher is pressurized by opening the valve (V-2) first, followed by valve (V-1) to pressurize the launcher and kick (launch) the scraper.
- Valve (V-3) is closed, to force the fluid to go through the launcher in order to push the scraper through the pipeline outside the launcher.
- Once the scraper is launched and passes through the valves, the launcher is isolated (V-3 is opened and V-1 closed followed by V-2).
- Launcher is drained in preparation for the next launching operation.

TABLE 6 What-If Analysis Form for Pipelines Pumps

What If?	Answer	Likelihood	Severity	Recommendations
Pump starts with discharge valve closed	• Increased discharge pressure leading to pump damage or pipeline rupture	Low	High	• Use pressure protection systems to shut down the pump • Implement pump start-up procedure to ensure valve is checked to be open before starting the pump • Monitor discharge valve position at the control room if possible
Level at storage tank decreases	• Pump can run dry causing cavitation that could damage the pump or causes fire due to heated fluid at the pump cavity or electrostatic charges formation	Low	High	• Use level control to ensure that the level at the storage tank is maintained above the critical limits • Use pump protection systems to initiate pump shutdown should the level drops below acceptable limits
Discharge pressure increases significantly	• Causes pump damage or pipeline rupture	Medium	High	• Use pressure protection systems to shut down the pump • Use pressure indicator to monitor the pressure at the pump discharge and link it to high-pressure alarm at the control room
Pump temperature increases	• Can cause pump damage and/or fire	Medium	Medium	• Use pump protection systems to initiate pump shutdown upon high-temperature detection • Use temperature *indicator to monitor pump temperature* and link it to high-temperature alarm at the control room
Flow increases	• Cause operational problems and could compromise the pipeline integrity on the long run	High	Low	• Use pump process control system to control the flow within acceptable limits • Monitor flow from the pump in the control room

(Continued)

TABLE 6 What-If Analysis Form for Pipelines Pumps—cont'd

What If?	Answer	Likelihood	Severity	Recommendations
Vibration increases significantly	• Can cause damage to pump and pipeline leading to a leak of hazardous material	Low	High	• Install vibration control systems and use pump protection system to shut down the pump if the vibration increases beyond acceptable limit
Pump starts with suction valve closed	• Causes pump cavitation due to the lack of fluid feeding the pump • This could damage the pump	Medium	Medium	• Use pump protection system to initiate pump shutdown in case the temperature of the pump increases as a result of this • Implement pump start-up procedure to ensure that valve is checked to be open before starting the pump • Monitor suction valve position and flow from the pump in the control room if possible

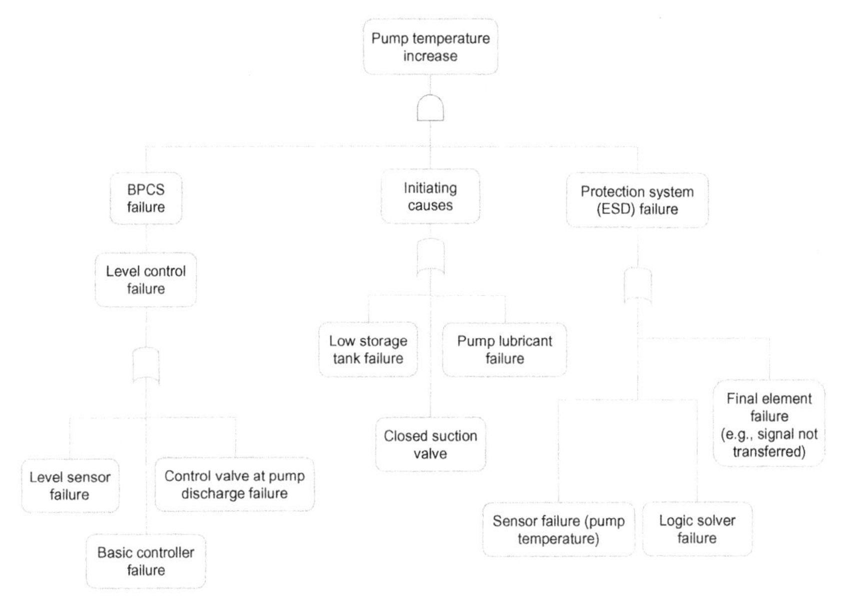

FIG. 12 FTA example of pipeline pump temperature increase.

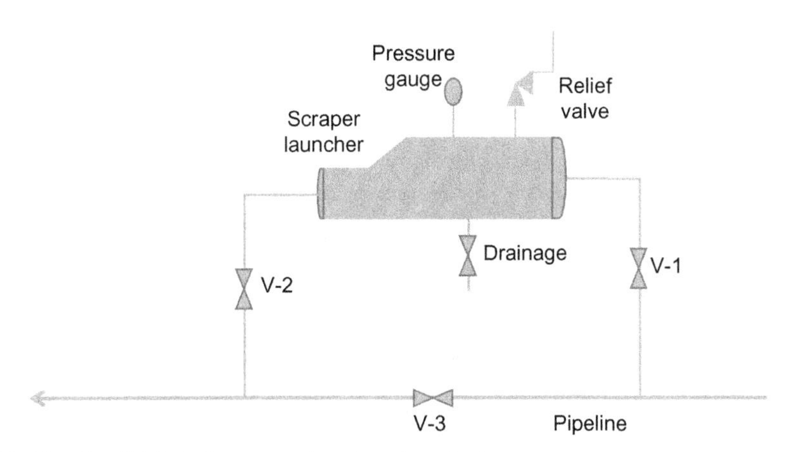

FIG. 13 Simplified pipeline scraper launcher schematic diagram.

Detailed procedure can be found in the literature, but this is a brief description in general.

As summarized in the scraper launcher procedure given above, there are different opportunities to cause incidents if the procedure is not followed adequately. For example, the following situation could lead to incidents and must be assessed during the HAZID study:

- If the launcher is opened before it is fully depressurized, this will lead to the loss of pressurized hazardous containment and will cause damage. This can be mitigated by following the clear/well-prepared procedures and by checking the pressure indicator/gauge to ensure that the pressure is zero before opening the door. Also, to prevent door breaking loose and launching at the operator, a door interlock mechanism should be in place to hold the door and prevent it from hitting the operator in case the launcher is opened while it is pressurized.

- If the scraper launcher is not pressurized enough after the scraper is launched, this could mean that the scraper could get stuck in the system and operation will not be complete. This might cause damage to the scraper, lost production, and wasted operator time as well.
- If the size of the scraper is larger than the maximum size that can be handled by the pipeline system or if dirt accumulates inside the pipeline to restrict scraper movement, the scraper could get stuck leading to flow restriction, overpressurization of the system, and damage to scraper itself.

As seen from these examples, what-if analysis can be conducted to ensure that hazards associated with scraper launching (and receiving) operations are assessed and addressed appropriately. Other techniques such as HAZOP could also be used to ensure that the design of the scraper launcher/receiver is appropriate by asking questions related to flow, pressure, etc., such as no flow, reverse flow, less flow, and high pressure.

Other Pipeline System Components

HAZID techniques can be used to assess hazards associated with the design and operation of the other components of the pipeline systems such as the storage tanks, metering stations, and chemical injection systems used to inject corrosion inhibitors and other chemicals. As demonstrated earlier in this chapter, HAZID studies will improve the design of the process and mitigate the risk associated with the operation of the pipeline and other components. A thorough assessment is needed, and it can be conducted following the basic concepts and approaches described in this chapter.

SUMMARY

This chapter has presented a detailed review of common HAZID techniques used in the process industry and demonstrated their application to pipeline situations. As was demonstrated, HAZID techniques complement each other and sometimes lead to the same conclusions/outcome. Some techniques are more common than others, and some are used at different stages of the life cycle of the pipeline. Table 7 summarizes the characteristics of different techniques and their potential versus limitations. More information is available in the literature [13–18].

HAZID assessments can be performed using software and computer packages, which are available in the market. However, as demonstrated in this book, a simple sheet can be used. This book is not recommending any software in particular. The user can review available software and choose the most suitable one.

TABLE 7 Summary of HAZID Techniques

HAZID Technique	Approach	Life-Cycle Phase	Resources	Skills Needed	Time Needed	Level of Information Needed
PHA	Brainstorming	Early stages	Team with basic information about process and hazard	Medium	Short	Basic pipeline information and service
HAZOP	Guide word structured approach	Design stage	Skilled team with experience in process design and HAZID techniques	Advanced	Long	Detailed P&ID drawings and process conditions, material and energy balances
What-if	Brainstorming with open Q&A session	Early stage and design stage	Experienced team with detailed knowledge of the design and operation of the facility	Medium-advanced	Short	Some details about process design and operational modes, as well as P&IDs if available
Bow tie	Cause-and-effect type of approach	Any stage	Experienced team with knowledge in HAZID and risk mitigation	Advanced	Medium	Detailed information about hazard available and process conditions
FTA	Mathematical approach for establishing causes and effects	Throughout the life cycle of facility	Experienced team with advanced knowledge in HAZID and strong operation/design background	Very advanced	Long	All information related to process from P&IDs, to design and operation information
ETA	Mathematical approach for establishing logical outcome of events following different potential paths	Throughout the life cycle of facility	Experienced team with advanced knowledge in HAZID and strong operation/design background	Very advanced	Long	All information related to process hazards and operational modes and process design

REFERENCES

[1] Preliminary Hazard Analysis Objectives, www.chambers.com.au.
[2] D.P. Nolan, Safety and Security Review for the Process Industries, Application of HAZOP, PHA, What-If and SVA Reviews, third ed., GPP, 2012.
[3] State of New South Wales—Department of Planning, Hazardous Industry Planning Advisory, Paper No 8, HAZOP Guidelines, 2011.
[4] L. Guo, J. Kang, An extended HAZOP analysis approach with dynamic fault tree, J. Loss Prev. Process Ind. 38 (2015) 224–232.
[5] P. Baybutt, A critique of the Hazard and Operability (HAZOP) study, J. Loss Prev. Process Ind. 33 (2015) 52–58.
[6] C.A. Ericson, Hazard Analysis Techniques for System Safety, Wiley, 2015.
[7] MIT Course Material on "Chemical Engineering Processes Laboratory," course 10.27, Appendix, VI http://web.mit.edu/10.27/www/, 1999.
[8] NIST, Fault Tree Analysis, https://csrc.nist.gov/Glossary, 2017.
[9] E. Ruijters, M. Stoelinga, Fault tree analysis: a survey of the state-of-the-art in modeling, analysis and tools, Comput. Sci. Rev. 15 (2015) 29–62.
[10] Online Search, Event Tree Analysis, http://www.eventtreeanalysis.com, 2017.
[11] N. Alileche, D. Olivier, L. Estel, V. Cozzani, Analysis of domino effect in the process industry using the event tree method, Saf. Sci. 97 (2017) 10–19.
[12] N. Alileche, D. Olivier, L. Estel, V. Cozzani, Analysis of domino effect in the process industry using the event tree method, Saf. Sci. 97 (August 2017) 10–19.
[13] N. Bariha, I.M. Mishra, V.C. Srivastava, Hazard analysis of failure of natural gas and petroleum gas pipelines, J. Loss Prev. Process Ind. 40 (2016) 217–226.
[14] I. Cameron, et al., Process hazard analysis, hazard identification and scenario definition: are the conventional tools sufficient, or should and can we do much better? Process Saf. Environ. Prot. (2017).
[15] S. Basu, Plant Hazard Analysis and Safety Instrumentation Systems, Elsevier, 2017.
[16] P. Baybutt, Requirements for improved process hazard analysis (PHA) methods, J. Loss Prev. Process Ind. 32 (2014) 182–191.
[17] R. Singh, Pipeline Integrity Handbook, second ed., GPP, 2017.
[18] T.A. Kletz, HAZOP & HAZAN, Hazard Workshop Modules, IChemE Information Exchange Scheme, 1986.

Chapter 3

Introduction to Pipeline Risk Assessments

Risk is defined in different ways depending on people's understanding and utilization of the term. The dictionary Merriam-Webster has different definitions of risk; the most relevant of them states that risk is "the possibility that something bad or unpleasant (such as an injury or a loss) will happen." Technically, risk has a precise meaning that is in line with the definition above but more specific. Technically, risk is defined as "a combined measure of potential damage in terms of the magnitude of the damage (consequences) and the likelihood (frequency) that this damage will occur."

Damage is defined as the type of damage that can occur to the following:

- Humans: injuries and/or loss of life
- Environment: damage to sensitive environment such as marine life and underground water reservoirs
- Assets: financial losses due to damage of pipeline facilities and/or public assets and loss of production and spill cleaning cost

The consequences are defined as the magnitude of damage resulting from a given event (e.g., leaks of hazardous material from the pipeline otherwise known as loss-of-containment (LOC) incident) should this event occur, regardless of the likelihood that this event will occur. This is not to be confused with the term "hazard" that is defined as the potential to cause harm, which will only materialize when being exposed to the hazard source. So, it is important to differentiate between these different terms (hazard, consequence, and risk) as they have different unique definitions and implications from risk management point of view.

For pipelines transferring flammable or toxic material, the following hazards exist:

- Flammable/combustible hazards: associated with the flammable nature of material that can cause fire upon ignition, such as the following:
  - Pool fires resulting from ignition of combustible or liquid flammable material
  - Jet fires resulting from ignition of pressurized gas or volatile flammable material such as liquefied natural gas (LNG) or liquefied petroleum gas (LPG)
  - Flash fires and explosions resulting from late ignition of flammable clouds in open and confined/congested areas, respectively

Cross Country Pipeline Risk Assessments and Mitigation Strategies
https://doi.org/10.1016/B978-0-12-816007-7.00003-2

- Toxic exposure resulting from release of toxic material such as ammonia or hydrogen sulfide to the atmosphere

In this book, the focus will be on risk to people from loss-of-containment events associated with pipeline failure.

PIPELINE RISK MANAGEMENT APPROACHES

Pipeline risk management can be addressed using one of the two approaches described below.

Prescriptive Approach

This approach is based on well-established rules and guidelines that dictate certain measures/actions to be implemented in a given situation in order to control/mitigate the risk. It focuses mainly on the consequence side of the risk management process and does not always account for how likely the event (situation) will occur.

For example, the prescriptive approach would mandate a minimum separation distance between the pipelines and public facilities regardless of the mitigation measures used to control the risk of the pipeline (i.e., exclusion zone that could be defined based on pipeline size not magnitude of risk). This exclusion zone could be outside the potential hazardous zone or immediate risk zone (e.g., outside flammable limit zones resulting from the leak of hydrocarbons from the pipeline). Pipeline classification also represents some sort of prescriptive approach, as it mandates minimum requirement for pipeline design factor and wall thickness based on population count within a certain hazardous zone from pipelines.

This approach is effective in controlling hazard and risk from pipeline but can sometimes be expensive or impractical, especially when it comes to maintaining exclusion zones around pipelines or considerably increasing wall thickness for long segments of the pipeline. The prescriptive approach is typically enforced through a set of standards, guidelines, regulations, and best practices.

Risk-Based Approach

This approach takes into account the probability of an event happening besides its consequences. It also credits the mitigation measures that can be utilized to control/reduce risk by either lowering its likelihood of occurrence or mitigating its consequences once it occurs. This approach is convenient in situations where the prescriptive approach requirements can't be met. However, a clear criteria and conditions must be available to establish limits of acceptable risk and respond to emergencies should they occur.

The risk-based approach offers a systematic methodology to identify hazards, assess the associated risks, and adopt risk control measures to manage pipeline risks. The process illustrated in Fig. 1 is typically followed to manage risks associated with pipeline operations and activities on people, environment, and assets.

Risk management process for pipelines is described in detail in this book. The following terminologies apply:

- Hazard evaluation: in this step of the risk management process, all types and sources of hazards in the systems being evaluated are identified in a systematic manner to ensure the hazards are fully identified and accounted for.
- Risk analysis: in this step of the process, the consequence and frequencies (likelihood) of occurrence are estimated for each hazardous event identified in the hazard evaluation step above. Then, these results of consequences and frequencies are combined together to estimate/calculate the risk values. Risk values of all hazards are combined together to predict the total risk levels of the systems being evaluated.
- Risk evaluation: the risk levels calculated above are compared with the tolerance/acceptance criteria to evaluate whether the risk levels are acceptable or not.

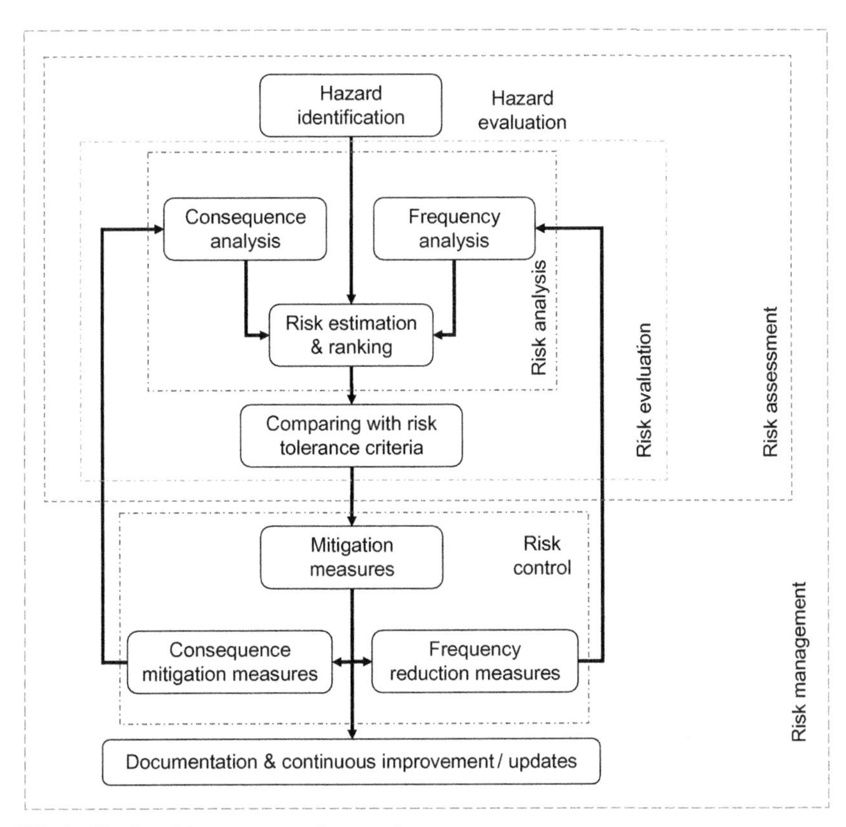

FIG. 1 Pipeline risk management framework.

Risk criteria have to be defined and accepted to be used. Further discussion on the risk acceptance criteria is given later in this book.
- Risk assessment: this step of the risk management process includes all activities conducted in the different steps above.
- Risk control: if the risk is not tolerable/acceptable, then risk mitigation and prevention measures are defined and evaluated to reduce the risk to acceptable levels per the tolerance criteria.
- Risk management: once the risk levels reach an acceptable/tolerable limit, then the process is documented for monitoring and continuous improvement to ensure the risk levels remain manageable. The whole process mentioned here from hazard evaluation to risk documentations and continuous improvement constitutes the full risk management process.

DIFFERENT TYPES OF RISK ASSESSMENT & MANAGEMENT STUDIES

There are two main approaches that can be used for pipeline risk assessments: qualitative and quantitative approaches. Each approach has its merits, limitations, and purposes. A summary of these approaches is given in the following sections.

Qualitative Risk Assessments

Risk is assessed qualitatively by using discrete classes or ranking the frequency and consequence associated with each type/source of hazard (e.g., high/medium/low). Risk is then defined using a risk matrix to rank or estimate risk levels. A simple risk register can be used to document qualitative risk assessment outcomes, similar to the one shown in Table 1. This will be illustrated in the next chapter.

At a minimum, a risk shall be assessed qualitatively by a multidisciplinary team with relevant experience to estimate, rank, and develop practical controls of the risk. A qualitative risk assessment is typically conducted for simple systems or activities that require relatively inexpensive risk controls, such as the following examples:

- Early stages of facility design, where process information is not detailed or formal
- Simple facilities where there is no impact on surrounding population or facilities (e.g., pipelines not passing through populated areas)
- Simple activities where interactions between different parties are not present

While qualitative risk assessments are simple and relatively quick to implement, they still have their limitations, such as the following:

- The approach is not suitable to address highly complex systems or a complex/chain of initiating events.

TABLE 1 Risk Register Table Used in Qualitative Risk Assessment

Type of Hazard	Cause	Consequence	Initial Risk			Mitigation	Recommendation	Residual Risk			Action Owner
			L	S	R			L	S	R	
Flammable release											
Toxic release											
Others											

- It is not intended for determining the cumulative risk of a facility.
- This approach can be subjective, and its success would therefore depend on the knowledge and experience of the team conducting the study, making it vulnerable to inconsistencies and interpretation of the team members.

Quantitative Risk Assessments

Risk is assessed quantitatively using detailed frequency and consequence analysis, where data are available and where the complexity of the system requires such details. Quantitative risk assessment (QRA) utilizes a systematic mathematical approach to predict the risk in terms of consequences and frequencies of hazard sources. Typically, results of the QRA are compared with established risk tolerance criteria or with risk results of other alternatives to evaluate options with lower risk.

Quantitative risk results are typically presented in different formats:

- Individual risk (IR): this is the amount of risk experienced by a single individual over a given time period. It is presented in numbers (individual risk per annum, IRPA) or as risk contours plotted on a map of the site. Risk contours assume the person is present all the time and don't account for presence factors (i.e., portion of the time the person is not present in that location). IRPA accounts for the time an individual is not present in at the risky locations and therefore is more representative of the actual risk experienced by a specific individual.
- Societal risk (SR): this is a measure of the total risk endured by a group of people and depends on the values of IR and the total number of people that are present in that location. It can be presented on an F-N curve (frequency of fatalities vs. number of fatalities) or as potential loss of life (PLL).

The detailed work required to conduct QRAs is typically intensive, and they are conducted only if a qualitative approach will not produce reliable results. QRAs should be used for evaluating alternatives when risk reduction measures are costly or when high-risk scenarios are identified. The following are some examples where QRAs are conducted:

- Complex processing facilities where aggregation of major risks is needed to determine design options. The availability of accurate and detailed process information (e.g., PFDs, P&IDs, plot plans, population information, and surrounding facilities) is needed to conduct QRAs.
- High-risk or complex operations where interactions between different groups are present.
- If the risk reduction measures identified by qualitative assessments are cost prohibitive or impractical.
- When evaluating different options of operations or facilities where qualitative risk assessment cannot produce detailed results for decision-making.
- When evaluating emerging technologies that cannot be assessed qualitatively.

- Severe events, nonroutine tasks, or potential risk to third parties and the public.
- A major accident in the pipeline industry that suggests the need for additional safety measures.

QRAs are typically expensive and time-consuming and may not be effective when data are insufficient or inaccurate. Also, the quantitative nature of the process may give a false sense of accuracy. The fact that the results of a QRA are expressed in numerical form may give an impression of greater accuracy than warranted, even if appropriate qualifying language is included. QRAs are described in details later in this book.

RISK ANALYSIS (CONSEQUENCES VS. LIKELIHOOD)

To estimate the risk associated with a facility or an activity, the consequence and frequency of the different hazardous events have to be analyzed. The risk will be then estimated and ranked. Risk analysis is conducted, in line with the process shown in Fig. 1, through the three steps shown in the following subsections.

Consequence Analysis

Consequence of an event is defined as the type and extent of hazardous impact resulting from that event. An event in this context typically refers to incidents (or accidents) that lead to hazardous situation, such as the leak of flammable material that ignites and exposes personnel to fire/thermal hazards leading to injuries or death. Consequence analysis is conducted to determine the impact and severity of the hazard (e.g., how many people are impacted and how many are likely to be injured or killed).

Frequency Analysis

In order to estimate the risk of a given activity or process/facility, the frequency (i.e., the likelihood) that different events will occur must be estimated and combined with the consequences of those events to get the risk values. Frequency analysis is usually a statistical analysis conducted based on data collected from the actual incidents occurring in the field. Different techniques are implemented such as the fault tree analysis (FTA), event tree analysis (ETA), or simple statistical data analysis.

For pipeline incidents, different databases are available for event frequencies such as Concawe for liquid pipelines leaks, EGIG for gas pipelines leaks, and OGP database for process equipment leak frequencies. Instrument failure frequency has their own databases as well, and those are used to estimate frequencies of safety instrumentation failure leading to process incidents. It is important to use a relevant database for estimating the frequencies, in order to have realistic/representative risk estimation.

In summary, the frequency analysis is conducted to determine the likelihood of occurrence of the hazardous events being evaluated. A detailed pipeline failure analysis is included in this book in Chapter 5.

Risk Estimation and Ranking

Risk is estimated by combining consequence and frequency of different hazards. The estimated risk of different events is aggregated to estimate the overall risk. Risks are typically estimated for personnel and public facilities around the pipeline facilities, the business losses, and the environment.

Risk contributors are then ranked based on their level/magnitude. This allows for allocating risk controls effectively by addressing the highest risk contributors in the facility or operation being assessed. If the risk assessment is conducted to compare different facilities, configurations, operations, or routes of the pipelines, then the ranking of the associated risks will help decide the optimum situations. As stated above, risks can be analyzed either qualitatively or quantitatively, and concepts described here are applicable in both cases.

Consequence vs. Likelihood

Since risk is the combination of consequences and likelihood, the risk can increase if either the consequences or the likelihood increases or both. As such, it is very important to understand that not every high-consequence event is a high-risk event, unless it is highly frequent (i.e., has high frequency). In order to verify that, let's consider the following example:

If someone decided to travel from City A to City B, he/she could either fly or drive. Flying is safe, but in the case of an airplane accident, the consequences could be quite severe. A single airplane incident could cause a couple of hundred fatalities. However, these incidents do not happen very often. Typically, only few airplane incidents occur every year causing low number of fatalities every year compared with the total numbers traveling by air worldwide. Given the large number of passengers flying every year, the likelihood of someone dying in an incident is very low (statistically speaking). This is an event with low frequency and high consequence. The risk is deemed low.

On the other hand, considering the large number of car accident-related and vehicle accident-related fatalities, the frequency of accidents on the roads is very high. In fact, there are much more fatalities every year from car accidents than from any other mean of transportation worldwide. However, car accidents typically result in a relatively low number of fatalities compared with airplane incidents (typically less than three). As such, this is a high frequency-low consequence event but still causing high risk.

It is clear from this example that a high-consequence event can be a low-risk event, while the low consequence-high frequency event can be high risk. This may not always be the case though. So, from risk point of view, it is recommended to fly from City A to City B rather than driving (Table 2).

TABLE 2 Consequence vs. Risk of Transportation

Activity	Event	Consequence	Frequency	Risk
Driving	Car accidents	Low	High	High
Flying	Airplane crash	High	Low	Low

RISK CRITERIA

In order to determine if the risk is acceptable or not, the risk results must be compared with an established criteria that define the acceptable levels of risk. Risk tolerance criteria (RTC) is often defined and dictated by regulatory requirements in terms of unacceptable and/or acceptable levels of risk resulting from a facility or an activity. RTC should explicitly define the levels of acceptable risk to be adopted when assessing risks from pipelines. The following three risk limits are typically defined by the RTC:

- Intolerable (unacceptable and high risk): risks that fall within the intolerable region cannot be justified and have to be reduced by employing effective risk control measures. Consideration should be given to stopping the risky operation or activity until the risk is reduced, if needed. This can be achieved by changing the design and operation or even rerouting the pipeline away from population centers and personnel to reduce the risk to acceptable levels.
- Broadly acceptable (relatively low risk): risks that fall within the broadly acceptable region should be maintained at low levels by employing adequate and cost-effective/obvious risk controls. Risks in this region are not a cause of major concern.
- Tolerable (medium risk): for risks that fall in the middle region between intolerable and broadly acceptable regions (i.e., tolerable risk), the as low as reasonably practicable (ALARP) principle can be followed. The ALARP principle allows evaluating the practicability (cost-effectiveness) of the different risk control measures. Most effective risk reduction measures are adopted to reduce risk and maintain it at low levels (as low as possible) without grossly investing in the reduction measures. This is called demonstration of the ALARP. Cost-benefit analysis can be implemented to demonstrate that the intent of the ALARP concept has been established.

These concepts are illustrated in the schematic shown in Fig. 2. Ideally, the risk should be as low as possible, but the lower the risk target is, the higher is the cost required to bring the risk to this target level. As such, the concept of ALARP should always be demonstrated in the case of high-risk operations.

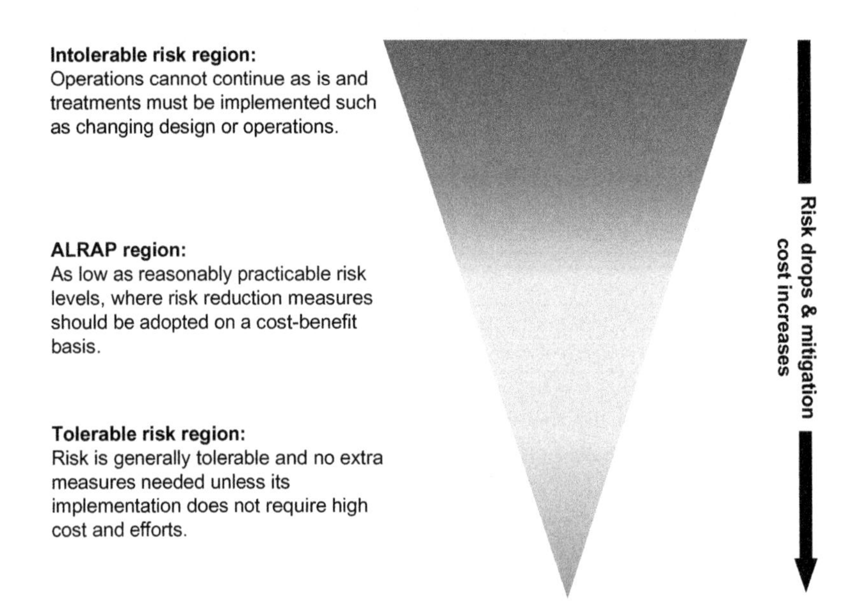

FIG. 2 ALARP concept.

RISK MATRIX

Qualitative risk tolerance criteria addresses the risks to people, environment, and business through the risk matrix. The risk matrix allows estimating the risk by selecting the likelihood (frequency) and severity (consequence) of the hazard. It also allows determining required risk reduction measures in a manner appropriate to the level of risk. A generic risk matrix that could be used in risk assessment studies is presented in Fig. 3, which is a "five-by-five" matrix. Other grids can be used based on the specific needs of the pipeline operators.

Tables 3 and 4 provide examples of generic definitions of the likelihood and severity levels required to estimate risks using the risk matrix. Table 3 provides guidance for frequency estimation for each identified hazard that will be based on historical data and engineering judgment. Table 4 provides guidance for consequence estimation of each identified hazard, in terms of impact on people, environment, assets, etc. that are estimated using engineering judgment and experience of the team performing the assessment (as these assessments are typically conducted by a team).

Once the frequency and consequence are analyzed, a risk level is obtained for each hazard, by plotting the frequency and consequence in the risk matrix. This risk can then be compared with risk criteria shown in Fig. 3, so that its tolerability can be judged. The matrix described here is shown for demonstration purposes mainly. It is not necessarily a standard matrix.

The table above can be expanded to include reputation, legal implications, and other relevant parameters/categories.

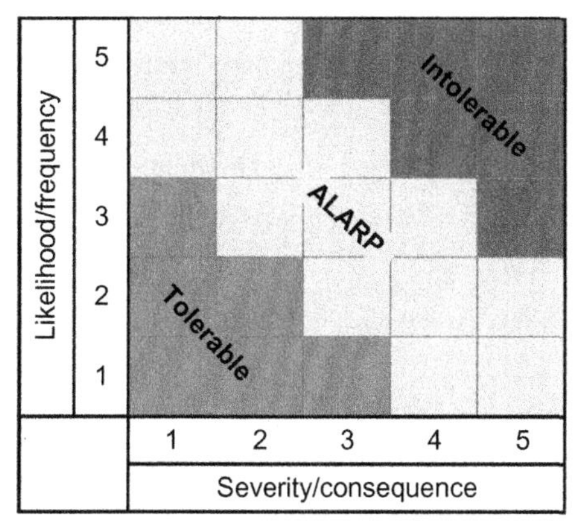

FIG. 3 General five-by-five risk matrix.

TABLE 3 Risk Matrix Likelihood/Frequency Categories

Likelihood Level	Explanation	Example
1	Practically not possible	Event that is theoretically possible but quite rare
2	One-of-its-kind event that is possible to happen but not heard of often	Event similar to Deepwater Horizon incident that is quite unique
3	Has happened before in pipeline operations	Large catastrophic release that could not be controlled for a long time
4	Happens multiple times in the pipeline being evaluated	Small release of material due to pinhole release, defected O-ring, or seal on a poster pump
5	Regular event	An event that is common and happening in the pipeline operations multiple times every year

RISK CONTROL AND MITIGATION

Following risk assessment and evaluation, necessary control/mitigation measures should be evaluated and adopted to manage the risks and maintain it within tolerable risk limits. Risk mitigation strategies typically follow the structured 4Ts, as outlined below:

- *Terminate risk completely*: measures should be evaluated to eliminate the risks if possible by adopting measures that remove the source of risk such as the inherently safer designs of the process or the use of separation distance, which eliminate the impact on the public and adjacent facilities. Maintaining

TABLE 4 Risk Matrix Severity/Consequence Categories

Severity Level	Explanation	Safety	Environment	Economic
1	Insignificant impact that requires effectively no intervention	Minor injury with first aid treatment on-site	Insignificant localized/ unmeasurable environmental effect	Insignificant impact that can be absorbed as a part of regular operational cost
2	Some impact that requires some resource to treat it	Major injury that requires treatment off-site	Local impact on the environment that lasts long	Low-cost impact that does not hinder or affect operations
3	Moderate impact on safety, environment, and operational business continuity	Serious injury leading to permanent disability	Large impact that can be rectified in reasonable time with reasonable effort (not affecting operations)	Temporary financial impact and loss of review that can be absorbed without affecting operators' financial obligations
4	Significant impact that can affect the ability to continue operation	Single fatalities	Significant widespread impact that lasts for some time and requires large effort to rectify	Significant financial impact that affects operator's bottom line and financial stand and lasts for a considerable duration
5	Severe impact that jeopardize the ability to continue operating the pipeline	Multiple fatalities or serious injuries	Long-lasting damage to the environment that requires massive remediation efforts	Severe cost impact or losses that can lead to shutting down the business or effectively driving the operator out of business

proper separation distance by routing the pipeline away from the population or environmentally sensitive areas can reduce the impact in case of incidents. Some of the other risk termination strategies can also help, such as manufacturing the hazardous material (e.g., chlorine and ammonia) close to the customer site, which can reduce the length of pipelines needed to transfer them or eliminate the need for the pipeline, and this removes risk sources altogether.

- *Treat the risk*: if the risk cannot be eliminated as described above, it should then be minimized by implementing risk control/mitigation measures, such as the following:
  - *Design and engineering measures*: two types of measures are available, designing prevention and mitigation measures. Prevention measures reduce the frequency of hazardous incidents occurring, while mitigation measures aim at reducing the consequence of the hazard incidents once they occur. For example, proper corrosion management and inspection programs will help reduce the frequency of leak from the pipeline, while reducing pressure, temperature, and pipeline size if possible can reduce the impact of the released material.
  - *Implementing administrative controls*: these controls include the development of operation procedures that are designed to minimize operational errors and human errors that can lead to incidents and increase the risk of the pipeline. Other administrative measures also exist and can be utilized to minimize the risk, such as the emergency response plans.
  - *Using appropriate personal protective equipment*: this helps to protect people in case of incidents and minimize the risk of getting injured or killed as a result of these incidents. For example, toxic masks can be effective against toxic releases, while fire-resistant clothing can protect against flash fire and thermal hazards.
- *Transfer risk to others*: if risk can be better managed by other specialized organizations, the operation or activity may be transferred to those parties. Insurance is an example of managing potential asset risks, by transferring risk/liability to the insurance company. Another example is related to scraping operations, which can be conducted by specialized contractors leading to potential reduction of human error and hence risk.
- *Tolerate low levels of risk*: if the risk levels are relatively low, they can be tolerated provided that they can be maintained at low levels. This is linked to the ALARP concept described above. ALARP must be established in order for the risk to be tolerated. The risk must be monitored and evaluated on a regular basis in order to make sure it remains within tolerable limits.

Detailed description of measures that can be used to improve pipeline integrity and reduce the risk will be described in Chapter 6 of this book.

Chapter 4

Pipeline Consequence Modeling

Consequences of pipeline incidents (i.e., leak from pipelines) are defined as the expected damage from exposure to the different type of hazards associated with the leak. For example, the consequence of loss of containment (LOC) incident (leak) from pipeline could be fire (jet, pool, or flash), blast, or toxic impact depending on the nature of the material. In order to be able to estimate the extent of potential damage (i.e., consequences) and how far it can reach, consequence modeling must be conducted. In addition, consequence modeling must be done in all risk assessment to calculate the risk values. Risk is a combination of both consequence and likelihood as was explained in Chapter 3. A detailed description of different types of consequence modeling and how they are performed is given in this chapter.

In order to perform consequence modeling, the analyst should first identify the potential hazard and potential outcomes from this hazard. A HAZID exercise should be conducted for that purpose, and then an event tree should be generated to establish the possible outcomes that need to be modeled as a result of a given initiating event as was demonstrated in Chapter 2.

LOSS OF CONTAINMENT (LOC) INCIDENTS

For incidents to occur, initiating causes must exist and act on the safe and normal operations to convert it into unsafe mode of operation and expose the surrounding to the hazardous consequences. Initiating causes can range from natural disasters to falling objects, external impact, corrosion, and incorrect operation. These are process-related and nonprocess-related causes and will be discussed in details in the next chapter. For pipelines, most of the process-related incidents are associated with LOC of either flammable or toxic material that can lead to further escalation of events (e.g., fires or explosion). This will be the focus of discussion in this chapter.

Many hazardous materials are transported using pipelines in large quantities either in liquid or gaseous forms. Pipeline failure could lead to release of the material contained to the surrounding atmosphere. The status of the released material and its quantities are determined by its physical properties and the operating condition (e.g., temperature and pressure). Accidental release of contained material can range from small release (resulting from pinholes) to large releases (resulting from ruptures of pipelines). The released material may form a liquid pool, disperse to the atmosphere directly (if it is released as gas),

Cross Country Pipeline Risk Assessments and Mitigation Strategies
https://doi.org/10.1016/B978-0-12-816007-7.00004-4

or ignite immediately (such as pure hydrogen, which has high tendency to ignite immediately upon release). Detailed mechanisms of different release incidents and formats are given somewhere else [1–14]. Fig. 1 provides details on the potential outcome of pipeline incidents based on the released material conditions and physical properties.

Fig. 2 defines the different steps needed to complete the consequence modeling of pipeline LOC incidents. Consequence modeling calculations involves the following steps:

- Selecting the LOC (i.e., release) scenario (typically selected based on the HAZID results).
- Estimating the discharge rate and conditions (source modeling). This includes determining the resulting phases (i.e., vapor, liquid, two phases, mist, and aerosols), temperature, and discharge velocity.

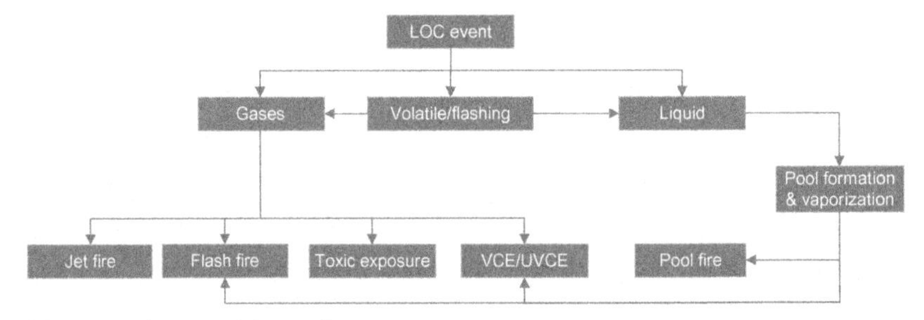

FIG. 1 Pipeline potential scenario outcomes.

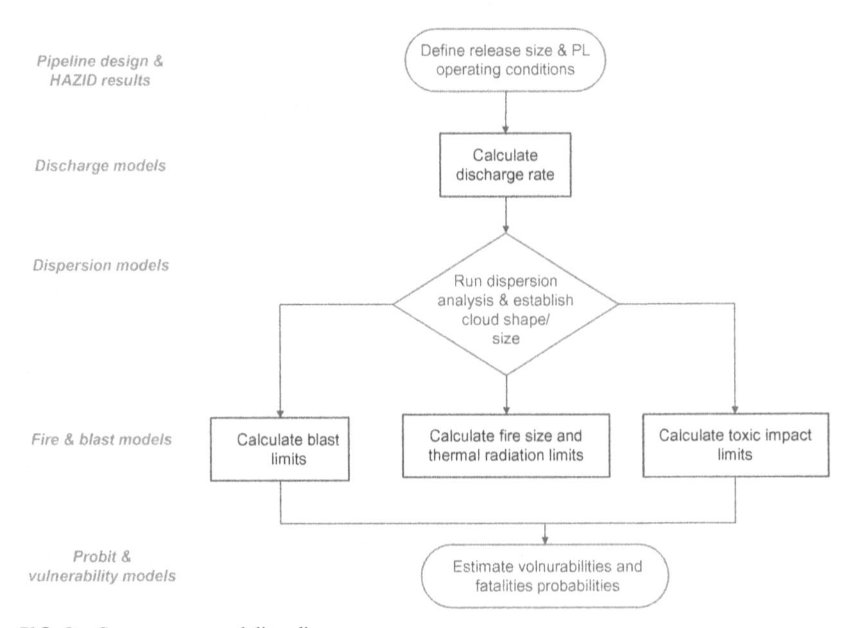

FIG. 2 Consequence modeling diagram.

- Conducting consequence modeling, which involves the following:
  - Dispersion of the released vapor or the vapor generated from a vaporizing liquid pool
  - Effect modeling, which includes assessing the extent and effects of the consequences identified:
    - Fires (pool, jet, and/or flash fires) resulting from ignition of the released flammable material.
    - Vapor cloud explosions (VCEs) resulting from an ignition of a vapor cloud released into a congested/confined area.
    - Unconfined vapor cloud explosions (UVCEs) resulting from an ignition of a vapor cloud without congestion or confinement. This typically occurs for highly reactive material.
  - Vulnerabilities evaluations, which calculates probabilities of fatalities due to different types of consequences

Causes for LOC can range from external impact to corrosion, operational upsets, and human errors. Statistical data show that there is a range of release sizes (pinhole, small/medium leaks, and large leaks) that are associated with these release mechanisms or failure modes of pipelines. The consequences of these releases are directly related to the release size, which causes different release quantities. Fig. 3 shows generally the different types of consequences that follow LOC incidents, which are common in the pipeline industry.

Initiating event	Material phase	Contains toxic	Early ignition	Late ignition	Congestion confinement	Consequence outcome
LOC	Gas	No	No	No		Gas release (pollution)
				Yes	No	Flash fire
					Yes	Explosion
			Yes			Jet fire
		Yes	No	No		Toxic exposure
				Yes	No	Flash fire
					Yes	Explosion
			Yes			Jet fire
	Liquid	No	No	No		Oil spill (pollution)
				Yes	No	Flash fire/pool fire
					Yes	Explosion
			Yes			Jet fire/pool fire
		Yes	No	No		Toxic exposure
				Yes	No	Flash fire
					Yes	Explosion
			Yes			Jet fire/pool fire

FIG. 3 General event tree for consequence modeling.

DISCHARGE CALCULATIONS

In order to model the consequences of pipeline leaks, the release rate resulting from the release must be quantified first. The released material quantities and the physical/thermodynamic characteristics will provide quantitative information that can be used in subsequent calculations to determine the impact of the material release from its containment inside the pipeline (i.e., consequence modeling). The amount and flowrate of material released depend mainly on the following factors:

- Process conditions, such as pressure and temperature inside the pipeline
- Physical properties of the material being transferred in the pipeline, such as material characteristics, mixture compositions, and the phase of material (which is evaluated using thermodynamic models)
- Pipeline design parameters, such as pipeline size and flowrates inside it
- Release size being modeled

Different discharge calculations (models) are available for different material types and conditions, which include the following:

- Nonflashing liquid
- Gases: subsonic and supersonic gases
- Flashing liquids

Detailed models are available in literature for all types of release conditions [1,2]. The models presented in this book are simplified versions of these models that produce acceptable results but in a simplified form that is suitable for pipeline risk assessments.

Liquid Discharge Model

The following equation describes the discharge rate model for liquid material [1], which can be used for calculating release rates associated with pipelines:

$$\dot{m} = 0.68\, D^2 \sqrt{\rho P_1}$$

where

$\dot{m}$: Liquid discharge rate (kg/s).
D: Release diameter (m).
ρ : Fluid density (kg/m^3).
P_1: Pressure (gauge) inside the pipeline (Pag).

As seen in the equation, the release size, material density, and pressure in the pipeline have significant impact on the calculations. This equation is based on formulas given in Ref. [1], with discharge coefficient of 0.6. It gives conservative estimates of the release rate. Detailed model can be used from the original reference and other literature [1,2].

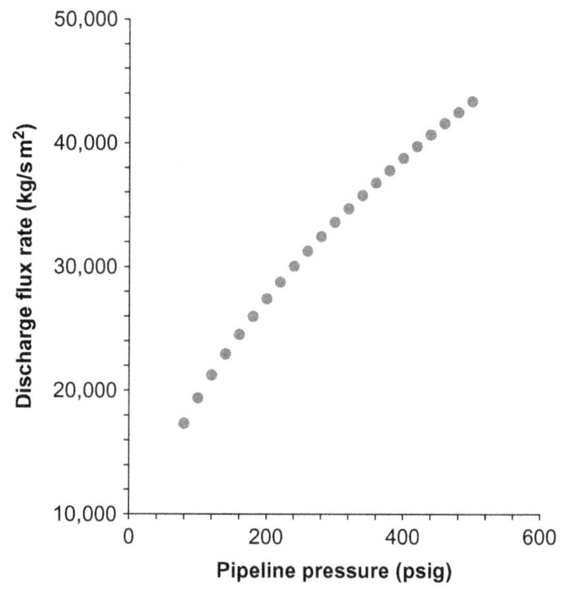

FIG. 4 Liquid gasoline pipeline discharge flux rates.

Fig. 4 shows the mass flux rate (kg/s m^2) versus pipeline pressure for liquid gasoline pipeline with the following conditions:

- Pipeline size = 20 in.
- Gasoline density = 740 kg/m^3

Fig. 5 shows the discharge rate calculation results for the same gasoline pipeline, as a function of pipeline pressure and release size. Figs. 4 and 5 are based on the detailed model calculations, but the formula above gives an estimate that is very close and conservative.

Gas Discharge Model

For gases, the calculations are more complicated and depend on whether the released gas flow is subsonic or sonic. It also depends on whether the flow will chock or not. Generally speaking, for typical cross-country, high-pressure pipelines, which are the scope of this book, flows will chock and can therefore be considered as sonic flows. This will be the case, as these pipelines have pressures that are characteristic of chocking flow, per the criteria defined in other references [1]. Based on the isothermal discharge model given in the literature, the following formula is developed to give conservative estimates of gas discharge rates for typical gas properties:

$$\dot{m} = \frac{D^2 P_1}{160}\sqrt{\frac{M}{T_1}}$$

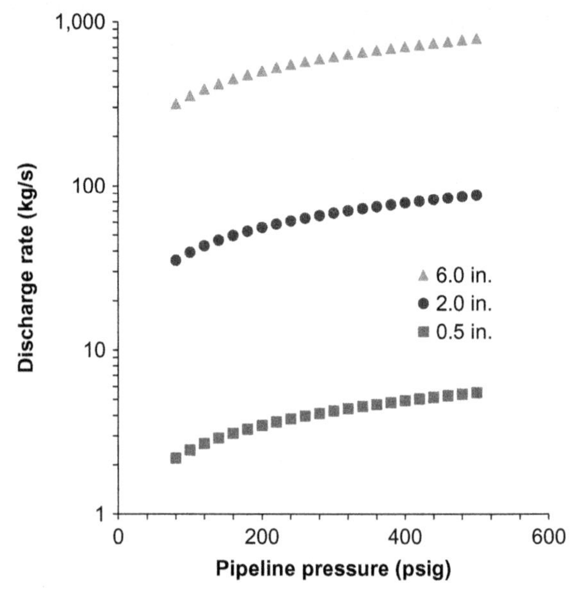

FIG. 5 Liquid gasoline pipeline discharge rates per release size.

where

$\dot{m}$: Gas discharge rate (kg/s).
D: Release diameter (m).
P_1: Pressure inside the pipeline (Pag or N/m^2 g).
M: Molecular weight of the gas (g/mol).
T_1: Gas temperature inside the pipeline (K).

This formula provides conservative estimates assuming discharge coefficient of 0.85. Detailed modeling can be found in the literature. Fig. 6 presents the gas discharge flux rate for natural gas (NG) versus pipeline pressure using the detailed modeling. The discharge mass flowrate is given in Fig. 7 for the same pipeline as a function of the pressure and release size.

Flashing Liquid Discharge Model

The discharge model for flashing liquid (volatile liquid) is more complicated and requires more details. A brief description of the model is given below, and detailed description of this model is given somewhere else [1,2]:

$$\dot{m} = A\sqrt{G_{sub}^2 + \frac{G_{erm}^2}{N}}$$

where

$\dot{m}$: Discharge rate (kg/s).
A: Release area (m^2).

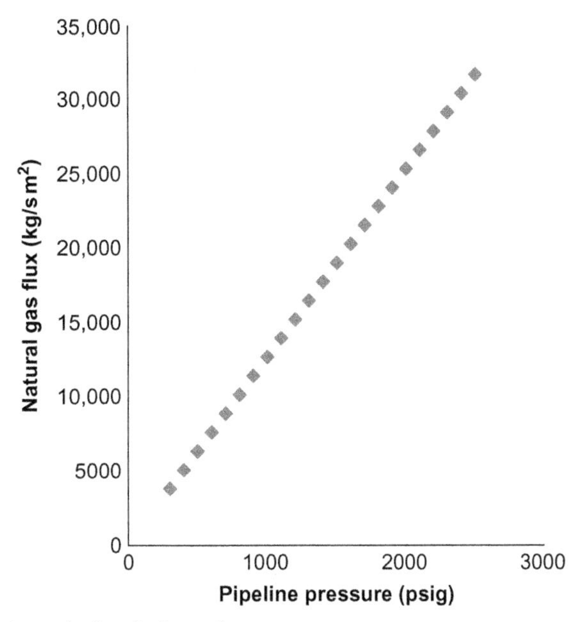

FIG. 6 Natural gas pipeline discharge flux rates.

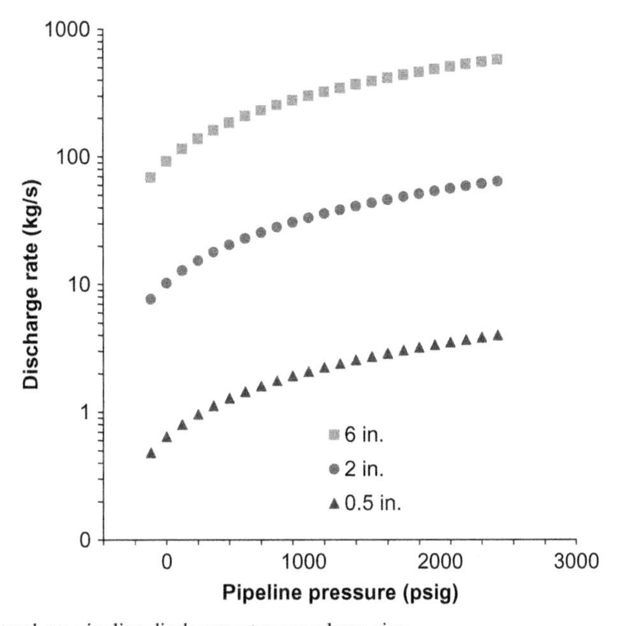

FIG. 7 Natural gas pipeline discharge rates per release size.

G_{sub}: Subcooled mass flux (kg/(s m^2)).
G_{erm}: Equilibrium mass flux (kg/(s m^2)).
N: Dimensionless nonequilibrium parameters.

However, for pipeline cases, the length of the release orifice is small (can be assumed to be close to the pipeline thickness). As such, the material discharging from the pipeline will not flash inside the orifice and will be discharged as liquid. So the liquid discharge model described earlier can be used to predict the release rate for flashing liquids. Computer models in commercial software packages include detailed calculations that can be used in the analysis. For calculations by hand, the subcooled model can be used as a conservative approach.

As seen from these discharge models, the pressure upstream of the release point, the physical properties of the material (e.g., density, heat capacity ratio, and thermodynamic state), and the release sizes determine the amount of discharged flowrate being released from the process/storage equipment.

It is important to note that the discharge models can predict unreasonably high flowrate for large release sizes of highly pressurized pipelines (i.e., 6.0 in.—full bore rupture). In such cases, the flowrate should be limited to a reasonable value. Some work was done by the author and concluded that a maximum flowrate of 3–5 times the normal flowrate (3*NFR–5*NFR) inside the pressurized pipeline can be reasonably conservative enough. So, if the discharge models predict flowrate higher than 3*NFR–5*NFR, the value should be replaced by the 3*NFR–5*NFR (depending on how conservative the analyst wants to be). However, if the risk analyst chooses to use other factors or approaches, then that should be fine as well, provided that there is a good justification for that. The information provided here is suggested only as a general guideline.

The normal flowrate is typically linked directly to the design velocity of the pipeline. For liquid pipelines, the optimum velocity ranges from 1 to 5 m/s, and for gases, the velocity ranges from 3 to 15 m/s [15]. These values can be used to estimate the normal flowrate in case such information is not available when assessing the discharge rates from pipelines for risk assessment purposes. If no details are available on the factors dictating the velocity (such as liquid and corrosive material presence), the upper range should be considered to be conservative. Fig. 8 shows an example of NFR for liquid gasoline pipeline as a function of pipeline size.

The mass fraction of the released volatile liquid that flashes into gases and disperses to the atmosphere is given in the following equation [1]:

$$F_v = C_p\left(\frac{T - T_b}{h_{fg}}\right)$$

where

C_p: Heat capacity of the liquid (averaged between T and T_b).
T: Initial temperature of the liquid.
T_b: Atmospheric boiling point of the liquid.

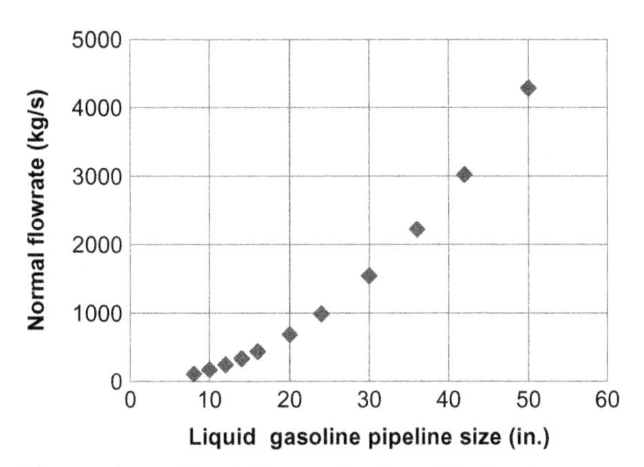

FIG. 8 Normal flowrate in gasoline pipeline as a function of pipeline size.

h_{fg}: Latent heat of vaporization for the liquid at T_b.
F_v: Mass fraction of released liquid that vaporizes.

The flashed gas can be treated as a released gas, while the liquid fraction will form a pool for consequence modeling calculations. For conservative risk assessment calculations, it is assumed that aerosols will be equivalent to the flashed gas rate and that all of it will evaporate and nothing will rain out [1].

Evaporation from the boiling pools and evaporating pools is estimated using evaporation rates due to heat transfer from the ground. Detailed models are available in open literature. The rates depend mainly on the pool diameter, time after pool formed, and material physical/thermal characteristic.

For the boiling liquid pool, evaporation is driven mainly by heat transfer from the soil, and the following formula can be used to estimate the evaporation rate:

$$\frac{\dot{m}}{A} = 0.0042\frac{\left(T_g - T\right)}{\left(t\right)^{0.5}}$$

where

$\dot{m}$: Evaporation rate (kg/s).
A: Pool area (m).
T_g: Soil temperature (°C).
T: Liquid pool temperature (°C).
t: Time after spill (s).

This formula is developed from detailed models available in the literature assuming that solar radiation will be typically negligible compared with thermal radiation from the ground at early stages of leaks. It will only be important in later times, and to be conservative, it is recommended to take early rates of evaporation for the calculations. The formula also uses the average soil type thermal diffusivity (0.2 mm^2/s) and thermal conductivity (in the order of

1.0 W/m K), as well as typical hydrocarbon latent heat of vaporization (in the order of 300 kJ/kg).

The pool area is needed to estimate the total mass flowrate. Pool area and diameter are given later in this chapter.

DISPERSION MODELING

Once the material leaks from the pipeline, gases and flashing material form an airborne cloud directly, while liquid release forms a pool that will eventually start vaporizing and form a cloud. The cloud will disperse into the atmosphere mainly due to turbulence and interaction with the wind. The profile of the dispersed cloud and its concentration (at different locations and times) can be estimated using the dispersion modeling calculations. As the cloud moves downstream of the release point (or source), it gets diluted through mixing with air, and its size increases to cover a wider area.

The main mechanisms of dispersion depend greatly on the nature and physical properties of the material being released and other ambient conditions. Mainly the mechanisms are dictated by air entrainment in early stages of the plum (near field) followed by diffusion in later stages of the plum (far field) [16–18]. The detailed description of dispersion mechanisms is well established and described in other Refs. [1,2,6,17–19]. Entrainment of the surrounding air is very important for cloud dispersion. The higher the entrainment rate is, the quicker the cloud dispersion will be.

Some of the parameters that are important include wind speed and direction relative to the release direction and the stability category of the wind (how turbulent wind is). Whether the material is buoyant or not will also play an important role in the dispersion process.

Dispersion modeling provides estimate of the downwind concentrations of the released material to a given end point (hazardous level). The level of hazard defined by the end point should be selected based on the level of damage that it can cause. For flammable material, typically, the end points are defined as the upper flammable limit (UFL), the lower flammable limit (LFL), and a concatenation equivalent to half of that ($^1/_2$ LFL). The rationale behind using these values is that they define the limits in which the cloud can be flammable and can ignite to cause fire and blast. One-half LFL is used to account for the nonideal mixing between the cloud and the air, where pockets rich with flammable material can exist outside the predicted LFL envelope (as mixing is not instantaneous between the cloud and the entrained air). For toxic end points, different levels of end points can be defined. In some cases, the end points might be based on the Emergency Response Planning Guidelines (ERPG) levels of the toxic material. These are defined based on the American Industrial Hygiene Association (AIHA) as shown in Table 1. In other cases, especially when doing risk assessment and to evaluate the vulnerabilities, different toxic end points are defined, which are associated with different levels of vulnerabilities such as the concentrations associated with 100%, 10%, and 1% fatality levels.

TABLE 1 Effects of Hydrogen Sulfide on People

ERPG Level	Effect on People
ERPG-1	The maximum airborne concentration below which it is believed that nearly all individuals could be exposed for up to 1 h without experiencing other than mild transient adverse health effects or perceiving a clearly defined objectionable odor
ERPG-2	The maximum airborne concentration below which it is believed that nearly all individuals could be exposed for up to 1 h without experiencing or developing irreversible or other serious health effects or symptoms that could impair their abilities to take protective action
ERPG-3	The maximum airborne concentration below which it is believed that nearly all individuals could be exposed for up to 1 h without experiencing or developing life-threatening health effects

There are many factors that affect the dispersion calculations of the released materials including those summarized in Fig. 9. These are discussed in sections below, but more detailed analysis is given in the open literature.

Weather Conditions

Wind speed has a significant impact on the cloud dispersion. As the wind speed increases, mixing with air slows it down, and the cloud will extend to further distance. However, atmospheric temperature, pressure, and relative humidity have a mild effect on dispersion, except for temperature effect on the evaporation rate from liquid pools. On the other hand, weather stability categories have a significant impact on the dispersion of the cloud.

The stability categories (often called Pasquill-Gifford categories) define the level of turbulent in the wind and are dependent on the general climate and solar radiation levels in the specific geographic location where the cloud

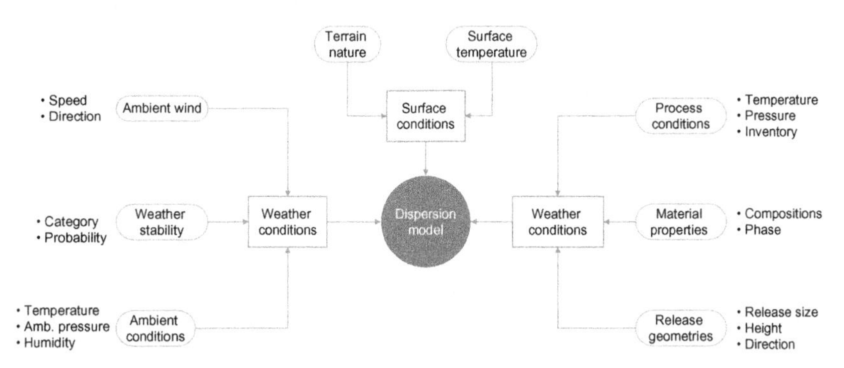

FIG. 9 Dispersion modeling parameters.

is dispersing, as well as the timing during the day or night. Generally, there are around five categories: stable, neutral, unstable, and couple of categories in between. Unstable wind represents strong turbulence conditions leading to quick dispersion of the cloud, while stable wind causes a reduced mixing rate between air and the cloud and slows down the dispersion of the cloud.

Table 2 can be used to estimate the weather categories under certain conditions [20]. Typically, weather data are collected over a long period of time (e.g., 3 years) and analyzed to get the averaged categories. However, if information is not available, dispersion analysis can be performed using two weather categories that represent day and night conditions. These are a combination of low wind speed (around 1.5–2 m/s) and stable weather category (called F category) and neutral weather categories (called D category) and higher wind speed (around 5.0 m/s).

The impact of the weather categories on the dispersion patterns of the cloud is illustrated in Fig. 10. The figure shows categories for stable and two categories for unstable weather conditions.

Surface Characteristics

The nature of the terrain represented by the parameter surface roughness can have an impact on the dispersion rate. Clouds that disperse on water or pen flat terrains will have lower mixing rates than clouds that disperse over forests and city centers with a lot of obstacles that will enhance turbulence and increase the mixing rate between air and the cloud. Surface roughness values range from 0.0001 m for smooth water surfaces to 3 m for city centers. Details of the surface roughness values can be found on open literature.

The surface temperature can impact rate of pool vaporization for liquid releases. The surface temperature will affect the rate of vaporization of the liquid and therefore the extent of the cloud.

TABLE 2 Estimate of Weather Category (Stability) Classes

	Day Time Solar Radiation			Night Time	
Wind Speed (m/s)	Strong	Moderate	Slight	Heavy Cloud Coverage	Low Cloud Coverage
<2	A	A–B	B	F	F
2–3	A–B	B	C	E	F
3–5	B	B–C	C	D	E
5–6	C	C–D	D	D	D
> 6	C	D	D	D	D

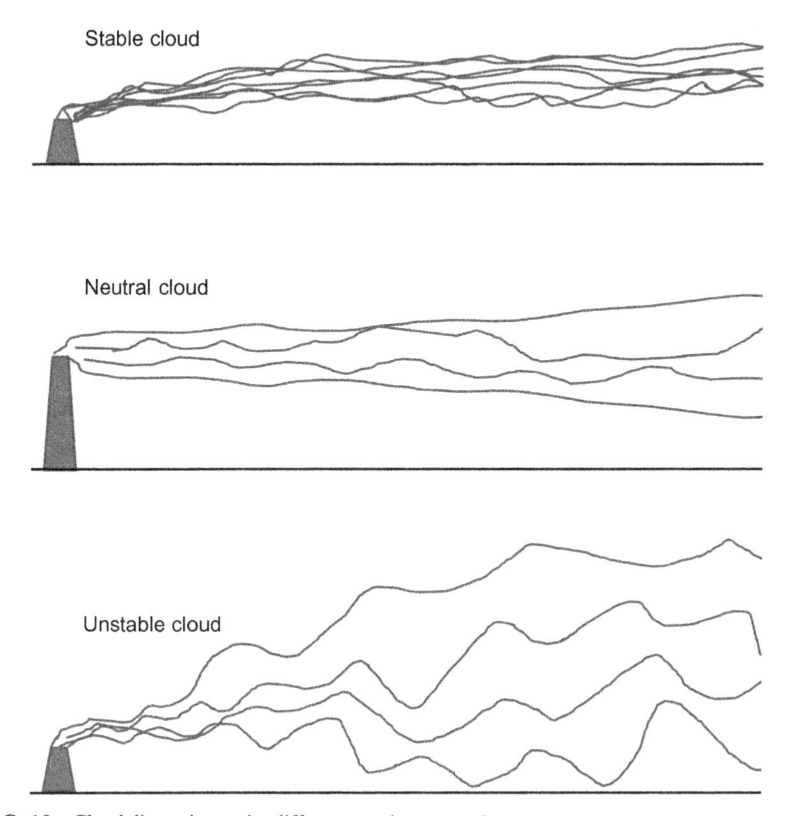

FIG. 10 Cloud dispersion under different weather categories.

Process Conditions

The temperature and pressure inside the pipeline will have an impact on the discharge rate (as was shown in the previous section) and on the physical properties of the released material. This will affect the dispersion rate as well. The higher the pressure inside the pipeline, the higher the release rate will be. High temperatures will increase the amount of the material flashing upon release or cause the material to evaporate faster.

The duration of the release and extent of the impacted area can depend on the inventory available in the pipeline. If the inventory is not isolated, a larger cloud can form with higher impact. Therefore, it is important to be able to isolate the inventory through emergency isolation valves along the pipeline or emergency shutdown systems (ESDs) in case of release.

Material Characteristic

The physical characteristics (such as density and compositions) affect the dispersion profile. Light material will have different dispersion profile and will cover a different area than the heavy materials. In general, light material will disperse generally faster than heavy gases and vapors. Also, material phase will have an impact. Gases and flashing vapors can behave differently compared with vapors from liquid pools.

Release Geometry

The release size and height are important as they will affect the discharge rate and interaction between the cloud and the ground. Larger releases will produce larger clouds, and elevated releases will have generally a chance to disperse to lower concentrations before the cloud reaches the ground. Release orientation will dictate how the cloud interacts with the surrounding wind and the surface, so vertical releases will be different dispersion rates and profile compared with horizontal one. Generally speaking, elevated vertical releases will disperse into the atmosphere before reaching the ground, and that will reduce their impact compared with horizontal releases close to the ground.

DISPERSION MODELS

Generally, the type of dispersion model selected depends on the cloud buoyancy and duration of release. There are several factors that determine the appropriate dispersion model to be used. Puff (short duration) versus plume (long continuous) models are available. Dispersion models should be able to handle the following types of problems:

- Jet dispersion analysis
- Buoyant (light) and dens (heavy) gas dispersion
- Multiphase/multicomponent, aerosol, or droplet modeling

However, for pipelines, the discussion will be limited in this book to continuous releases, assuming no credit can be taken for pipeline isolation, which is a common practice in pipeline risk assessments. The cloud can be positively buoyant, neutral, or negatively buoyant. Models are available for these different regimes. The description below is based on information available in main Refs. [1,2,5,6], where more details can be found. The Gaussian models are available for buoyant and neutrally buoyant cloud regimes, while dense cloud models can be used for heavily dense regimes (also called box models, such as the SLAB and Britter-McQuaid models).

Buoyant Clouds Dispersion Models

The Gaussian model is simple and easy to implement, but it cannot be used for heavy clouds (especially for large release cases). It also does not account for the near field portions of the cloud. It is more appropriate for the far field of portions of the cloud. So, it is more accurate for large releases but can still be used for small releases (buoyant and neutrally buoyant clouds as explained).

Dispersion models have been validated and are well developed. The Gaussian model specifically is well described and established [1]. A simplified form of the Gaussian model can be used for pipelines, since most of these pipelines are located on either the ground, below ground, or slightly above ground. Also assuming releases from pipelines occur in the same direction of the wind, which represents the worst case scenario, then the model can be simplified further by assuming the cloud is symmetrical around its center. With these assumptions, the Gaussian model reduced the following equation:

$$\langle C\rangle(x,0,0) = \frac{G}{\pi\sigma_y\sigma_z u}$$

where

$\langle C\rangle(x,0,0)$: Average concentration (mg/m^3).
G: Continuous release rate (kg/s).
σ_x, σ_y, and σ_z: Dispersion coefficients in the x, y, and z directions (m).
u: Wind speed (m/s).
y: Crosswind direction (m).
z: Distance above ground (m).

Values for σ_x, σ_y, and σ_z are given in other references, and their values depend greatly on the weather stabilities and surface roughness [1]. Fig. 11 shows the concentration profile for a release of methane at a rate of 5 kg/s under ambient temperature into the atmosphere at two different surface roughness conditions for weather category of F2 (F stability and 2 m/s). Fig. 12 shows the dispersion profiles for low surface roughness at F2 and D5 (D stability and 5 m/s). As shown in these figures, surface roughness and wind conditions have a significant impact on the cloud dispersion profiles, and hence, it is impacted area and risk levels.

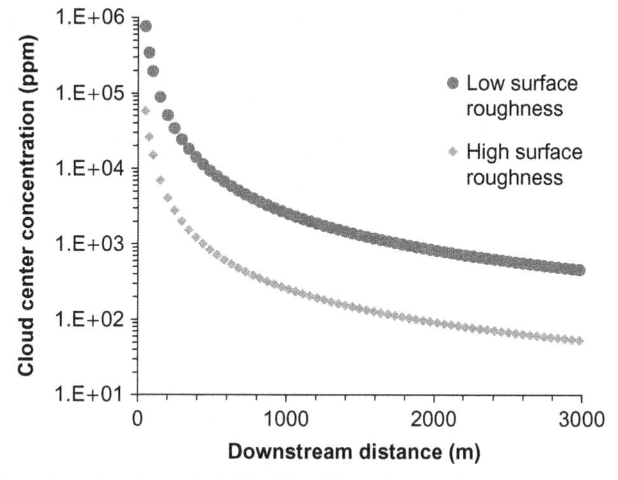

FIG. 11 Dispersion of methane cloud (5 kg/s) at F2 weather conditions.

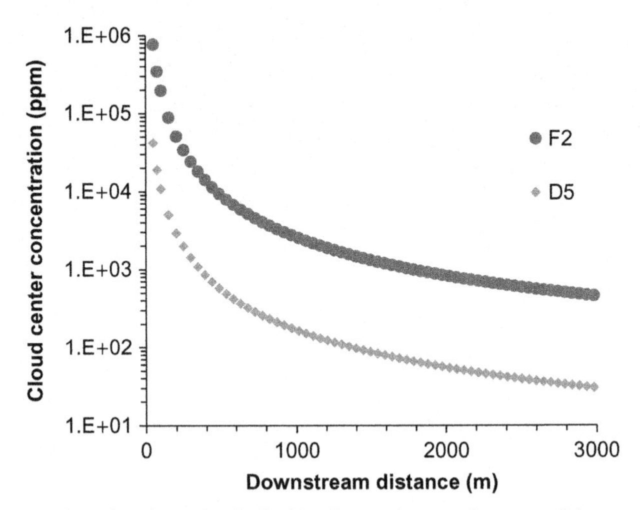

FIG. 12 Dispersion of methane cloud (5 kg/s) at low surface roughness conditions.

The analysis shown in these two figures assume an averaging time of 600 s. For flammable releases (no toxic included), the averaging time is much less (around 19 s). Toxic releases should use an averaging time of 600 s. The averaging time is an important parameter that defines the time needed to get a reliable average concentration in a given location inside the cloud that is stable enough to represent the concentration in that location.

The impact of the averaging time is given in the following formula:

$$\frac{\langle C(t1)\rangle}{\langle C(t2)\rangle} = \left(\frac{t2}{t1}\right)^{0.2}$$

where C is the concentration and t is the averaging time. As such, for flammable releases, the concentration at any location is almost double the concentration calculated with the toxic average time of 600 s. So, the curves shown in Figs. 11 and 12 will be higher by a factor of two since this is flammable release. Fig. 13 shows the concentration profiles for F2 low roughness factor methane release with 5 kg/s for averaging time of 600 s versus 19 s.

The model can also predict the shape of the cloud and the total coverage area of the cloud within a given concentration. For that, new parameters are defined [1]:

$$L^* = \left(\frac{Q_m}{u\langle C\rangle^*}\right)^{0.5}$$

$$x^* = \frac{x}{L^*}$$

$$A^* = \frac{A}{\left(L^*\right)^2}$$

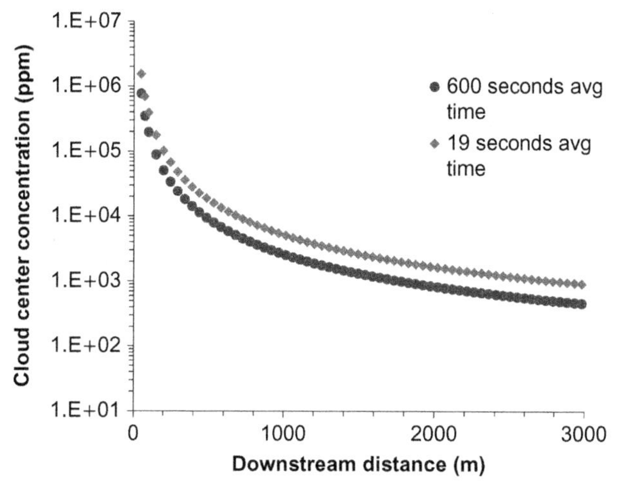

FIG. 13 Dispersion of methane cloud (5 kg/s) at low surface roughness for different averaging times.

where L^* is a scaled length, x^* is dimensionless downwind distance, and A^* is dimensionless area of the cloud. A^* and x^* parameters depend mainly on the wind stability class and are given in other references for different wind stability categories. Using these parameters, the cloud area and downwind distance to UFL, LFL, and ½ LFL are given in Fig. 14 for the methane release example mentioned above. These calculations are performed following the approach described in Ref. [1].

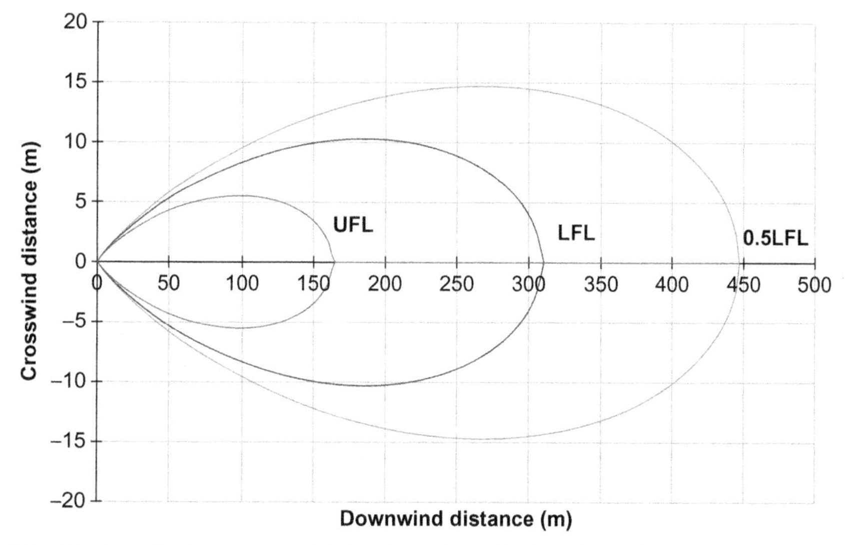

FIG. 14 Gaussian dispersion model of methane cloud (5 kg/s) at low surface roughness showing UFL, LFL, and ½ LFL at F2 weather conditions.

TABLE 3 Cloud Area and Maximum Width for Methane Example Shown in Fig. 14

Parameter	End Point	Value
Max. plume width (m)	0.5 LFL	14.7
	LFL	10.3
	UFL	5.5
Total area (m^2)	0.5 LFL	9627
	LFL	4663
	UFL	1320

Fig. 14 shows the distance to different end points of interest (UFL, LFL, and ½ LFL) for the methane example used here. Cloud maximum width and area are given in Table 3. All of these data are predicted by the model.

Dense Clouds Dispersion Models

Dense gas dispersion models should be used for dense clouds. One of the most simple dense gas models is the Britter-McQuaid model, which is well described and developed [1]. However, before using it, the cloud should be checked to confirm if it is dense or not. For continuous pipeline release, which is the scope of this book, the following equation should be satisfied:

$$\left(\frac{g_o q_o}{u^3 D_c}\right) \geq 0.003375$$

$$g_o = \frac{g\left(\rho_o \rho_a\right)}{\rho_a}$$

$$D_c = \left(\frac{q_o}{u}\right)^{0.5}$$

where

g_o: Initial buoyancy factor.
g: Gravity acceleration (m/s^2).
ρ_o: Density of released material in the pipeline.
ρ_a: Density of ambient air.
D_c: Characteristic source dimension (m).
q_o: Initial plume volume flux of dense gas dispersion (m^3/s).
u: Wind speed (m/s).

Using this model for *n*-butane release of 200 kg/s, the predicted distance to different flammable limits is given in Fig. 4. The predictions of the

TABLE 4 Comparison Between Predicted Distances for Different Dispersion Models

	Distance Predicted by Model (m)		
		Gaussian Model	
Limit	Britter-McQuaid Model	F2	D5
UFL	250	800	170
LFL	640	2200	420
½ LFL	1050	3550	640

Britter-McQuaid model (BMQ) are significantly lower than the values predicted by the Gaussian model for F2 weather conditions but higher than the D5 predictions of the Gaussian model. This could be due to the fact that BMQ model does not factor the surface roughness, averaging time, and weather category impact. As such, the analysis should make a decision whether to use the BMQ model or to be more conservative and use the Gaussian model (as it account for other parameters that the BMQ model does not account for). Not accounting for the correction factor for flammable, averaging time would have reduced the distances significantly (Table 4).

Pseudo Component Modeling

Simple dispersion models commonly used for risk assessments purposes, such as the ones described above, cannot handle mixtures of multiple components. So, a pseudomaterial is assumed to represent the mixture with one component for the purpose of calculations. The mixture components are used to calculate average properties based on the compositions fractions and that are used then as characteristics of the pseudomaterial. Dispersion analysis is then carried out similar to the single component case described earlier.

A special case is unique if the mixture contains both flammable and toxic components, where different end points have to be modeled for the toxic and flammable hazards with different averaging time. In that case, dispersion analysis must be conducted twice: once for the flammable cloud and once for the toxic cloud. However, the end point of flammable and toxic components will have to be normalized in terms of pseudomaterial concentration to accurately predict the impact and extent of these clouds. The following example illustrates that.

A mixture of gas contains 10% toxic H_2S material and 80% methane. The rest is inert N_2. In this case, the ERPG-2 and ERPG-3 of H_2S are 30 and 100 ppm. But since the mixture will be represented with pseudomaterial, then the dispersion will be conducted to new end points that are equivalent to 30 and 100 ppm of H_2S. Assuming ideal mixing and gases, then the 100 ppm H_2S will be equivalent to (100 ppm * 100 mol% pseudomaterial/10 mol% of H_2S), which will be 1000 ppm. The 30 ppm will be equivalent to 300 ppm of the pseudomixture. Same approach can be used for the flammable mixture as well. This is in essence the same approach used in commercial software as well.

FIRE AND THERMAL RADIATION MODELS

Fires occur as a result of the release of flammable material that can form a flammable mixture, which ignites by either an ignition source or through autoignition mechanisms. For a fire to occur, three components must exist together in the right format and conditions: oxygen, fuel, and an ignition source. However, the fuel must mix with air in the right ratio to form an ignitable mixture ready to burn. That ratio is defined as the flammable limits. For most flammable material, this ratio ranges from UFL to LFL as discussed earlier. The flammable limits are mainly physical properties of the material. Flammable limit significance is give below:

- UFL: the air available in the mixture is not sufficient to sustain the combustion reaction of material, and the fuel will not therefore burn.
- LFL: the fuel available in the mixture is not sufficient to sustain the burning, and there is too much air to allow a sustainable combustion reaction.

The autoignition temperature (AIT) is also another characteristic of the material that can determine whether a material burns immediately upon release or not. Table 5 shows some examples of autoignition temperature for some materials [21]. All pipeline leaks that occur at the material AIT or above it could ignite immediately.

There are three types of fires that can be modeled for pipelines carrying flammable material. These are flash, pool, and jet fires. For the purpose of this book, flash fires are represented with the flammable cloud boundaries of ½ LFL. Any person inside the flash fire is vulnerable, and anyone outside the limits of the flash fire is not. No extra models are specifically needed to assess the extent of flash fires. The discussion below will focus on pool and jet fires.

TABLE 5 Flammable Limits and Autoignition Temperature (AIT) for Common Materials

	Autoignition Temperature	
Fuel or Chemical	**(°C)**	**(°F)**
Benzene	560	1040
Butane	405	761
Carbon monoxide	609	1128
Charcoal	349	660

TABLE 5 Flammable Limits and Autoignition Temperature (AIT) for Common Materials—cont'd

	Autoignition Temperature	
Fuel or Chemical	(°C)	(°F)
Coal tar oil	580	1076
Diesel, Jet A-1	210	410
Ethane	515	959
Ethylene, ethene	450	842
Fuel oil no. 1	210	410
Fuel oil no. 2	256	494
Fuel oil no. 4	262	505
Gas oil	336	637
Gasoline, petrol	246–280	475–536
Heavy hydrocarbons	750	1382
Heptane	204	399
Hexane	223	433
Hydrogen	500	932
Isobutane	462	864
Kerosene	295	563
Light gas	600	1112
Light hydrocarbons	650	1202
Methane (natural gas)	580	1076
Naphtha	225	437
n-Butane	405	761
n-Heptane	215	419
n-Hexane	225	437
n-Octane	220	428
n-Pentane	260	500
n-Pentene	298	569
Petroleum	400	752
Propane	455	851
Toluene	535	995

Pool Fires

Pool fires result from the burning of flammable liquid pools. Typically, the impact of pool fires is localized and mild compared with other types of fires that can extend far from the source. The size of the pool has a significant impact on the damage that can result from pool fires. Pool fire size and extent depends mainly on the following factors:

- Released material and how quickly it can vaporize and burn.
- Pool size, which is the key parameter in determining the size/impact from pool fires. Limiting the size of potential pools (by dikes or proper drainage systems) can reduce the damage of pool fires
- Released inventory.

Pool fires are characterized by the following:

- Flame height above the pool
- Flame diameter
- Thermal radiation of the pool

Fig. 15 illustrates the pool fire dynamics. Pool fires typically spread vertically not horizontally, which limits their impact to the immediate areas nearby. To describe pool fires, it is first important to understand the pool formation mechanisms. A pool forms due to leak of liquid material or volatile material that flashes leaving liquid to form a pool and gas that disperse directly. Once the pool forms, it evaporates due to heat transfer from the ground or due to heat flux from the sun. If volatile material is released, the liquid forms will evaporate typically at a rate higher than the rate of evaporation from a total liquid pool.

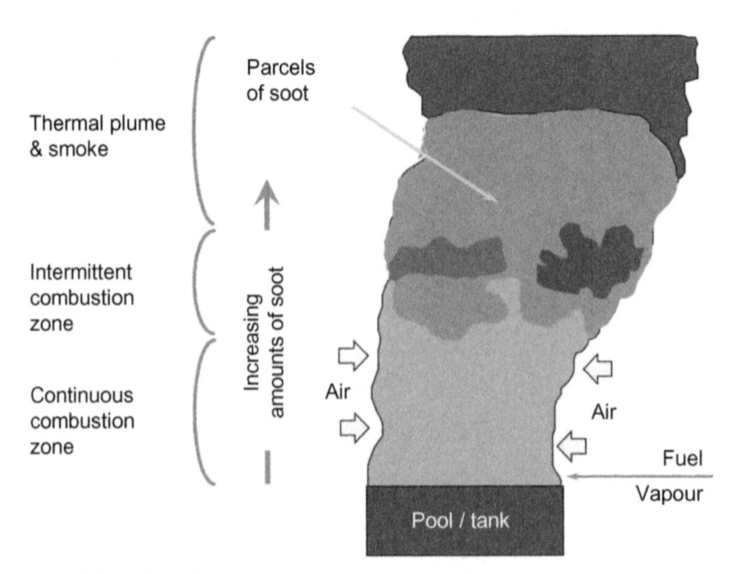

FIG. 15 Pool fire schematic.

The resulting pool in that case is referred to as boiling pool [1]. Flashed gases may also contain droplets of liquid that can evaporate or fall into the ground due to gravity. These are called aerosols. Detailed description of these mechanisms is given in the literature [6,19].

The pool diameter, flame height, and thermal radiation levels of the pool fire are defined by the following well-established models in the literature [1]. These models are simplified for the purpose of this book in to the following three formulas:

$$D_{max} = 135\sqrt{V_L}$$

$$H = 5.13D^{0.7}$$

$$X = \frac{10D}{\sqrt{E_r}}$$

where

D_{max}: Maximum possible pool diameter if pool is not restricted with a dike.
V_L: Liquid volumetric flowrate (m^3/s).
H: Flame height (m).
D: Pool diameter (m).
E_r: Thermal radiation level (kW/m^2).
X: Distance to reach E_r (m).

The above models are based on light/volatile hydrocarbons characteristics, in order to produce conservative results. Detailed analysis can be done using the original models. If the D_{max} is higher than the dike diameter, then D should be replaced with the dike diameter. Otherwise, it should be used as D_{max}. For pipelines, dikes may not be available most of the time. However, producing a large pool might not be realistic as well. So, it might be a good practice to limit D_{max} to an upper limit of 100m in reality; terrain nature around the pipelines would limit the growth of the pool. Table 6 gives an example for pool fire calculations using the above formulas with a dike diameter of 100m. It should be noted that these simplified calculations ignore the tilt and drag of the pool fire due to impact from the blowing wind, but since these models are conservative, this will not have a big impact on the results.

Jet Fires

Jet fires are the result of igniting pressurized flammable gas/volatile material at early stages of the release (i.e., close to the source of the release). Mainly, the jet fire extent is dependent on the following:

- Released material characteristics
  - Heating rate
  - Flame characteristics (e.g., adiabatic flame temp.)
  - Release orientation (vertical vs. horizontal)

TABLE 6 Pool Fire Calculation Example

V_L	D_{max}	D pool	H		X (m)	
(m^3/s)	(m)	(m)	(m)	37.5 kW/m^2	12.5 kW/m^2	4 kW/m^2
0.01	14	14	32	52	90	159
0.05	30	30	56	91	158	279
0.10	43	43	71	116	201	355
0.20	60	100	129	210	364	644
0.30	74	100	129	210	364	644
0.50	95	100	129	210	364	644
0.70	113	100	129	210	364	644
0.80	121	100	129	210	364	644
1.00	135	100	129	210	364	644

- Release rate
  - Momentum of the released jet at the release point
  - Energy released as a result of the combustion of the flammable material, which depends on the released rate and combustion heat

Jet fires are characterized by the following:

- Flame length
- Thermal radiation profile from the jet fire

Different models are available to calculate the jet fire such as the Chamberlain model and the Johnson model. Details are available somewhere else [22–24]. The flame length was estimated based on large-scale fire tests as given in the formula below [23]:

$$L = 2.89Q^{0.373}$$

where

L: Length of visible flame (m).
Q: Mas flowrate × the heat of combustion (MW).

Other formulas/models are available for the jet flame. Flame length for jet fire is given in a simplified model relative to diameter of the source in the following equation:

$$\frac{L}{d_j} = \frac{15}{C_T}\sqrt{\frac{M_a}{M_f}}$$

where

L: Length of visible flame (m).
d_j: Release size (equivalent diameter) (m).
C_T: Fuel mole fraction in stoichiometric fuel mixture with combustion air (unit less).
M_a: Molecular weight of air (g/mol).
M_f: Molecular weight of fuel (g/mol).

Thermal radiation flux received by a source from the jet fire is given by the simplified formula described below:

$$Er = 1200\frac{\dot{m}}{x^2}$$

where

Er: Radiation flux at the receiver location (kW/m^2).
$\dot{m}$: Mass flowrate of the fuel (kg/s).
x: Distance between point source and target.

This formula is developed from detailed models assuming

- typical weather conductions, with atmospheric *transitivity* of 0.75;
- maximum fraction of energy transferred into thermal radiation of 0.4 (for typical fuel);
- typical heat of combustion for common fuels to be in the order of 50,000 kJ/kg [25].

This simplified formula is likely to give conservative estimates of thermal radiation levels at the receiver location, but it is good for an estimate. More accurate ones can be evaluated using the original formulas. Fig. 16 shows thermal radiation profile for two releases of natural gas (NG) with release sizes of 6.0 and 2.0 in., predicted with the simplified formula.

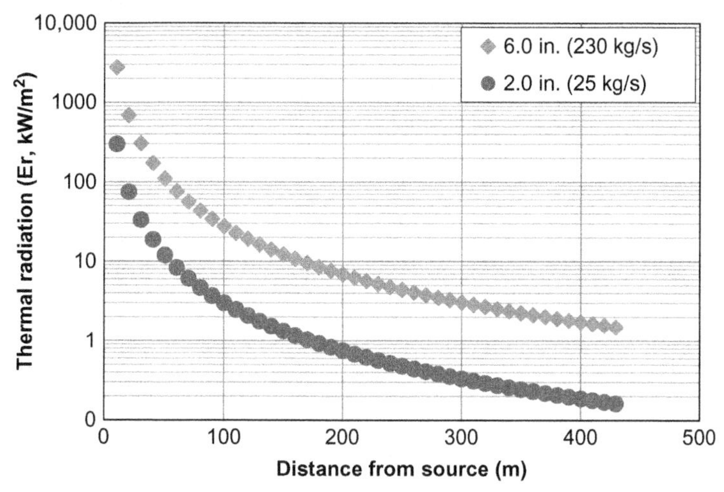

FIG. 16 Jet fire thermal radiation.

Thermal Radiation

Fires produce thermal radiation fields that can cause serious damage. Expected damage levels are summarized in Table 7. Vulnerabilities associated with each radiation level will be assigned to the thermal radiation produced by different fire types and will be reflected in the risk calculations.

BLEVE and VCE Calculations

Boiling liquid evaporating vapor explosion (BLEVE) is a special type of hazardous events that combines explosion with large fire. It mainly occurs when fire impinges on storage vessels of volatile light hydrocarbons vaporizing the content of the vessel and causes it to rupture leading to sudden rupture and release of the liquid with subsequent boiling and fire. BLEVEs tend to cause significant damage such as the event occurred in Mexico City in 1984 [26]. However, this event is unlikely to occur for pipelines and is therefore beyond the scope of this book.

VCEs would occur if a flammable cloud forms and intersects with a confined/congested area (such as process unit in plants and processing facilities) before it ignites in a late ignition event [2,10,27]. Typically, all flammable clouds that do not ignite early have the potential to cause VCE in process plants if the congestion/conferment is present. Unconfined flammable clouds for light and volatile materials tend to disperse quickly, and the probability of forming an unconfined VCE (UVCE) is therefore limited [10,27]. This is definitely the case for gases and light hydrocarbons. Heavy liquids do not tend to form flammable clouds [13,14]. If the clouds ignite in open areas, they form a flash fire, but not a VCE or UVCE, and that would be the typical case for pipelines outside the plant fence being discussed in this book. There have been cases where forest trees

TABLE 7 Thermal Radiation Levels

Radiation Intensity (kW/m^2)	Observed Effect (World Bank)
37.5	Sufficient to cause damage to process equipment
25.0	Minimum energy required to ignite wood at indefinitely long exposures (nonpiloted)
12.5	Minimum energy required for piloted ignition of wood, melting of plastic tubing
9.5	Pain threshold reached after 8 s can cause second degree burns after 20 s
4.0	Sufficient to cause pain to personnel if unable to reach cover within 20 s. However, blistering of the skin (second degree burns) is likely
1.6	Will cause no discomfort for long exposure

provided enough congestion to form a VCE, such as the incident in Buncefield (the United Kingdom, 2005). If this case exists for pipelines, then careful assessment of the explosion effect has to be done, following a simple model such as the TNT model for example. TNT model is well described in the literature and is not included in the scope of this book. Details of VCE and BLEVE are available in open literature [1,8,11,27].

UVCE is expected if large amount of HC is released (>20t) [28]. Using the TNT model and assuming the cloud takes a shape of a cone for simplification, the following procedure is used to estimate the UVCE impact:

- Estimate the size of the cloud based on cone volume:

$$\text{Volume}\left(m^3\right) = \pi r^2 \frac{h}{3}$$

where.

r is cone diameter (maximum width of the cloud divided by two).
h is cone height ($^1/_2$ LFL distance of the cloud).

Note that if the cloud is dispersing at the ground, then the volume would be half of the volume given in the formula above.

- The volume of the cloud is multiplied by the concentration of the flammable corresponding to LFL limit, which gives the volume of flammable material inside the cloud. This is converted into tons of flammables by multiplying it with the HC vapor density.
- The tons of HC are converted to equivalent TNT mass in tons (each TNT ton is equivalent to around 3t of HC) [28].
- Using information presented in Ref. [28], the distance to a given blast overpressure is given by the following formula:

$$D_p = 31.5 \sqrt[3]{\frac{\text{TNT tons}}{P^2}}$$

where
D_p is distance to a given pressure (m),
TNT tons is mass of TNT equivalent to HC released in tons,
P is overpressure from UVCE (bar).

- Assuming that the explosion takes place at the center of the cloud within the LFL zone, the distance to LFL is divided by two and is added to the D_p value given in the equation above. So, the distance to fatal overpressure levels resulting from UVCE (0.4bar) is therefore given by the following simplified formula:

$$D_p = \frac{\text{Distance to LFL}}{2} + 31.5 \sqrt[3]{\frac{\text{TNT tons}}{0.4^2}}$$

This is just a simplified formula. Detailed models are given in the literature as indicated earlier.

To illustrate this based on the dispersion example given in Fig. 14, the following calculations are performed using this model (Table 8):

TABLE 8 Distance to Lethal Overpressure Limit Example Calculations

Parameter	Value	Unit
Plum width	15	m
LFL distance	310	m
1/2 LFL distance	450	m
Cone volume (based on 1/2 LFL limits)	1.1E+05	m^3
NG LFL concentration	5%	mol%
NG 1/2 LFL concentration	2.50%	mol%
Flammable gas volume in the cloud	2.7E+03	m^3
NG vapor density	0.66	kg/m^3
NG tons	1.7	ton NG
TNT equivalent	0.6	Equiv. TNT ton
Pressure lethal limit	0.4	bar_g
D_p (distance to lethal pressure limits)	203	m

Based on these calculations, any person within 203 m from the pipeline is expected to receive fatal loads of overpressure from UVCE.

For buried pipelines, an "explosion"-like event takes place in case of pipeline rupture leading to the formation of a crater, which could resemble an explosion that leads to serious damage. This typically occurs for large leaks or pipeline ruptures, but not from small leaks. The width of the crater, where the damage is expected to be significant, is determined through several models such as the Advantica model, which determines the width of the crater as [12,29]

$$C_w \approx 0.4 D_p + 5.5$$

where

C_w: Crater width (m).
D_p: Pipeline diameter (in.).

This formula gives a good conservative estimate for the crater width up to pipeline pressure of 150 bar_g. The effect zone on each side of the pipeline is half of this distance. Anyone close to this zone should be assumed to be fatally injured for the purpose of risk assessment. For conservative estimate, a factor of 2 (i.e., twice the distance) can be used.[1]

1. This is just proposed to have a conservative estimate of the impact zone.

TABLE 9 Vulnerability Values for Thermal Radiation

	Thermal Radiation (kW/m^2) for Exposure Time (Seconds)			
Fatality %	**10 s**	**30 s**	**60 s**	**100 s**
1%	40	18	10	7
50%	90	40	21	14
100%	120	60	32	21

EFFECT MODELING

The purpose of this section is to define the percentage of potential fatalities that can result from the exposure to different levels of hazards. In this book, two types of hazards are considered: Flammable hazards leading to thermal impact only (since VCE and BLEVE are not considered credible), and toxic impact. Table 9 below shows thermal radiation limits that result in a given fatality probabilities (vulnerabilities) based on Mudan (1984) formula. For the purpose of this book and to be conservative, it is recommended to assume 100 s exposure time. This will allow personnel time to escape fires.

For flash fire, one can assume that everyone within flash fire envelope will have low chance of survival, and anyone outside it will have a high chance of surviving. For that, one can assume a vulnerability limit of 100% above LFL, 50% within ½ LFL, and 0 outside the ½ LFL. These values are just used for guidance only. If the analyst wishes to use other values, it should be fine provided that the values are reasonable and can be justified.

Toxic vulnerability can be evaluated using the probit formulas. Common toxic material transferred in cross-country pipelines includes H_2S, chlorine, CO_2, and ammonia. For these materials, the probit equations are given in multiple references. For oil and gas industry, the most common toxic material is H_2S. For this material, the probit equations are summarized below.

For toxic impact, it is recommended to use 60 min for impact on personnel from pipelines, which allows for enough time to alert affected personnel and for them to take measures to protect themselves. If it is anticipated that more time is needed, then appropriate values should be used accordingly. The figure below shows vulnerability data for different toxics based on US coast guard probit values (Table 10).

CONSEQUENCE MODELING SUMMARY

In summary, consequence modeling for pipeline is based on the models and formulas described in the literature [1–27,29–37] and is summarized by the chart shown in Fig. 17.

TABLE 10 Fatality Probabilities of Toxic Material Based on US Coast Guard Probits[a]

Material	Probability of Death	Exposure Time	
		30 min	60 min
H_2S	1%	256	158
	10%	327	202
	90%	594	366
Chlorine	1%	71	50
	10%	125	88
	90%	502	355
NH_3	1%	6147	4347
	10%	8164	5773
	90%	16,308	11,531

[a] *Concentrations shown in the table are in ppm.*

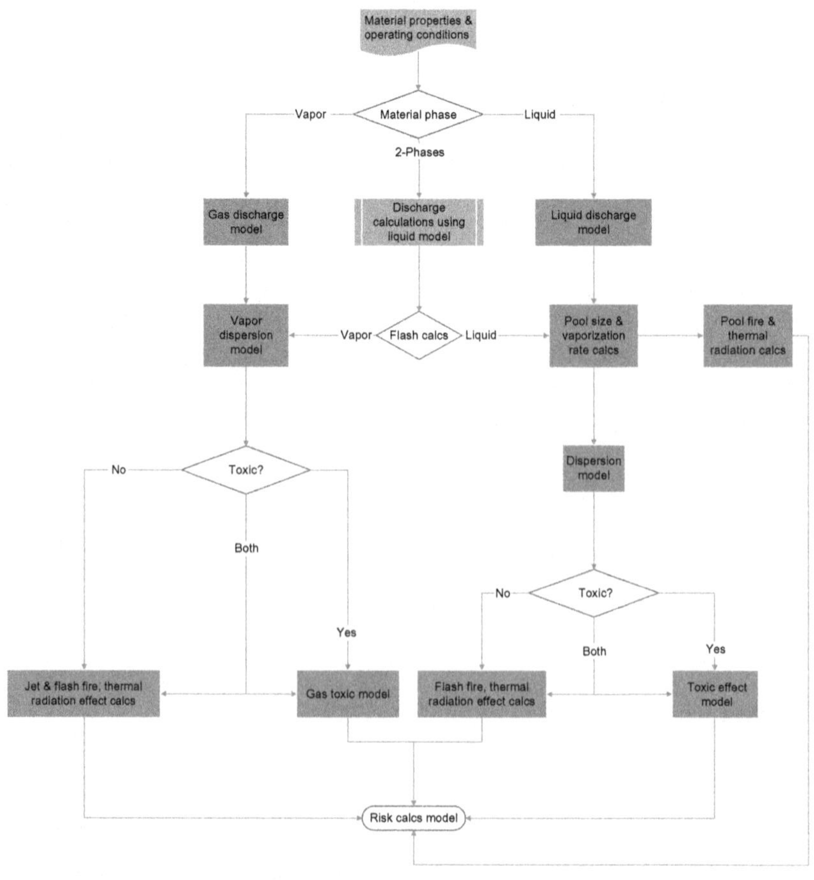

FIG. 17 Pipeline consequence modeling summary chart.

REFERENCES

[1] CCPS, Guidelines for Chemical Process Quantitative Risk Analysis, second ed., Wiley, New York, 2000.

[2] TNO Yellow Book, Methods for the Calculation of Physical Effects Due to Releases of Hazardous Materials (Liquids and Gases), third ed., 2005.

[3] TNO Purple Book, Guidelines for Quantitative Risk Assessment, first ed., 1999.

[4] TNO Green Book, Methods for the Determination of Possible Damage to People and Objects Resulting from Releases of Hazardous Materials, first ed., 1992.

[5] S.R. Hanna, P.J. Drives, Guidelines for Use of Vapor Cloud Dispersion Models, AIChE, New York, 1987.

[6] D.W. Johnson, J.L. Woodward, RELEASE: A Model with Data to Predict Aerosol Rainout in Accidental Releases, Wiley, New York, 1998.

[7] M. Parvini, E. Gharagouzlou, Gas leakage consequence modeling for buried gas pipelines, J. Loss Prev. Process Ind. 37 (2015) 110–118.

[8] O.R. Hansen, M.T. Kjellander, Potential for major explosions from crude oil pipeline releases in varied terrain, Chem. Eng. Trans. 48 (2016).

[9] H.W.M. Witlox, Overview of Consequence Modelling in the Hazard Assessment Package Phast, 2010, DNV, London, UK.

[10] P.L. Metropoloi, A.E.P. Brown, Natural gas pipeline accident consequence analysis, in: Oil and Hydrocarbon Spills III, WIT Press, Southampton, UK, 2002.

[11] N.S. Arunraj, J. Maiti, A methodology for overall consequence modeling in chemical industry, J. Hazard. Mater. 169 (1) (2009) 556–574.

[12] E.P. Silva, et al., Underground parallel pipelines domino effect: an analysis based on pipeline crater models and historical accidents, J. Loss Prev. Process Ind. 43 (2016) 315–331.

[13] W. Kent Muhlbauer, Pipeline Risk Management Manual: Ideas, Techniques, and Resources, third ed., GPP, 2004.

[14] J.L. Woodward, R. Pitbaldo, LNG Risk Based Safety: Modeling and Consequence Analysis, Wiley, 2010.

[15] Website on pipeline design, http://petrowiki.org/Pipeline_design_consideration_and_standards#Liquid_line_sizing.

[16] A. Aloqaily, Interaction Between Two Non-parallel Jets in a Confined Circular Duct, MSC thesis, University of Toronto, 2004.

[17] A. Aloqaily, A Study of Aerodynamics in Rotary Kilns with Two Burners, PhD thesis, University of Toronto, 2008.

[18] A. Aloqaily, et al., Effect of burning NCG on lime kiln flame patterns, J. Pulp Pap. Sci. 31 (2005).

[19] A. Aloqaily, Characterization of two-Phase Free Jet Flow, Master thesis, Department of Fluid Mechanics, Technical University Hamburg-Harburg, 2001.

[20] NAOO Website: www.ready.noaa.gov/READYpgclass.php.

[21] Engineering Tool Box Website: https://www.engineeringtoolbox.com/ (16.12.17).

[22] A. Aloqaily, A. Chakrabarty, in: Jet flame length and thermal radiation: evaluation with CFD simulations, AIChE Annual Meeting, 2010 Conference Proceedings, 2010.

[23] B.J. Lowesmith, G. Hankinson, M.R. Acton, G. Chamberlain, An overview of the nature of hydrocarbon jet fire hazards in the oil and gas industry and a simplified approach to assessing the hazards, Trans. IChemE Part B 85 (B3) (2007) 207–220.

[24] I. Coccorulloa, P. Russob, in: Jet fire consequence modeling for high-pressure gas pipelines, AIP Conference Proceedings 1790, December, 2016.

[25] Engineering Tool Box Website: https://www.engineeringtoolbox.com/ (16.12.17).
[26] A. Aloqaily, in: Industrial disaster management systems—lessons learned, Presentation at Asia Process Safety Summit, Malaysia, 2014.
[27] B. Rouge, in: Fundamentals of fires and explosions, AIChE SACHE Faculty Workshop, September, 2003.
[28] Tripartite Symposium, Assessment of Major Hazards in the Process Industries—Review of Current Methods, University of South Wales, 1985. November.
[29] M. Parvini, E. Gharagouzlou, Gas leakage consequence modeling for buried gas pipelines, J. Loss Prev. Process Ind. 37 (2015) 110–118.
[30] A. Chakrabarty, A. Aloqaily, in: Using CFD to assist facilities comply with thermal hazard regulations such as New API RP-752 recommendations, IChemE Hazards XXII, Symposium Series No. 156, 2011.
[31] https://jherring.files.wordpress.com/2012/05/img-2011.jpg.
[32] R.M. Peekema, Causes of natural gas pipeline explosive ruptures, J. Pipeline Syst. Eng. Pract. 4 (1) (2013).
[33] B. Rothwell, M. Stephens, in: Risk analysis of sweet natural gas pipelines: benchmarking simple consequence models, International Pipeline Conference, September 25–29, Calgary, Alberta, 2006.
[34] R. Alzbutas, et al., Risk and uncertainty analysis of gas pipeline failure and gas combustion consequence, Stoch. Env. Res. Risk A. 28 (6) (2014) 1431–1446.
[35] C. Davis, R. Williamson, Modeling pipeline spill determines impact on HCAs, Oil Gas J. 101 (12) (2003) 72–77.
[36] S. Sklavounos, F. Rigas, Estimation of safety distances in the vicinity of fuel gas pipelines, J. Loss Prev. Process Ind. 19 (1) (2006) 24–31.
[37] B.J. Lowesmith, G. Hankinson, Large scale experiments to study fires following the rupture of high pressure pipelines conveying natural gas and natural gas/hydrogen mixtures, Process Saf. Environ. Prot. 91 (1) (2013) 101–111.

FURTHER READING

[38] A. Rusin, K. Stolecka, Modelling the effects of failure of pipelines transporting hydrogen, Chem. Process. Eng. 32 (2) (2011) 117–134.

Chapter 5

Pipeline Failure Mode and Frequency

Pipelines are widely considered a safe mode for transporting hazardous material, and the pipeline network has been consistently growing in size throughout the world in the last few decades. Maintaining this vast network of important assets presents a serious challenge to owners of these assets and all other stakeholders involved including the local communities and authorities. The main issue that consistently manifests itself is the lack of credible data that allow for proper evaluation of potential release of hazardous material from these pipelines.

Loss of containment (LOC) of materials from the pipelines is the main events that should be presented and controlled to maintain the safety and protect the environment of the communities where these pipelines run. Pipeline failure modes and integrity assessment is a critical component in maintaining pipeline safety and managing its risk. This cannot be more critical than when these pipelines carry flammable/toxic material commonly processed/produced from chemical, petrochemical, oil, and gas facilities.

In this chapter, the potential modes and causes of failure of pipelines used in the oil and gas industry are evaluated based on a wide range of data available from different databases in the world that cover >40 years of operating history. The objective of the analysis is to develop a consistent approach that allows for proper estimation of potential failure frequency based on the actual conditions of the pipeline and using actual data without speculation. This will allow for proper assessment of the risk posed by the pipeline and estimation of the effectiveness of the mitigation measures used to control the risk.

Since the risk is a combination of the frequency of LOC events happening and the potential damage caused by these events, then the assessment presented in this chapter becomes critical in calculating the risk and mitigating it:

$$\text{Risk} = fr \times Co$$

where fr is the frequency and Co is the consequence in terms of potential injury and environmental damage or financial losses.

This chapter focuses on developing the fr estimation methodology. A detailed approach is needed as "not all pipelines are created equal." Pipeline failure modes and causes depend on several factors:

Cross Country Pipeline Risk Assessments and Mitigation Strategies
https://doi.org/10.1016/B978-0-12-816007-7.00005-6

- Design parameters, including size, wall thickness, coating type, burial depth, and backfill type
- Operating parameters, including temperature, pressure, and flowrate versus design rate
- Service conditions including type and phase of material, water content, and sour (H_2S and CO_2) content that can affect corrosion rates

Risk assessments need proper and accurate evaluation of pipeline failure frequency:

- Causes and modes
- Release size distribution
- Ignition probabilities

The lack of proper tool for detecting accurate pipeline failure modes and assessing its integrity affects the accuracy of the risk assessments. This leads to improper risk calculations and produces inaccurate/ineffective mitigation measures. In this chapter, a detailed analysis is presented to evaluate failure modes and causes for pipelines based on work conducted by the author and presented in Process Safety meetings.[1] This analysis was used to develop a new tool, which is also described in this chapter to assess failure modes and frequencies as well.

EXISTING APPROACH FOR EVALUATING PIPELINE FAILURES

Several specific databases that contain information about specific category of pipelines are available. These databases contain detailed information about the failure modes/causes of the specific categories of the pipelines. The main databases are as follows:

- *EGIG*: European Gas Pipeline Incident Data Group, 1970–2013
- *CONCAWE*: European Cross-Country Oil Pipelines, 1971–2013
- *UKOPA*: United Kingdom Onshore Pipeline Operators' Association, 1962–2013
- *US DOT PHMSA*: Pipeline and Hazardous Materials Safety Administration (PHMSA), 1995–2014
- *Canadian Database*: Canadian National Energy Board Report, 2000–08, and Canadian Pipeline Leaks Data, 2004–14

As seen from the description of the databases, each one focuses on a given category of pipelines. However, the US PHMSA databases really cover a wide range of categories, but the data are not readily available in the format that can be directly used in risk assessments for pipelines. The data need to be processed first to make them available in the needed format.

In this section, a detailed description of each of these databases is given to illustrate information available and scope of applicability/utilization of these databases.

1. MKOPSC work.

EGIG Database: 1970–2013

This database [1] covers the European gas transmission system pipelines that meet the following conditions:

- Clean and dry natural gas pipelines, which indicate that internal corrosion should not be a big contributor to overall frequency of leak.
- Collects data for onshore steel pipelines only with a maximum allowable operating pressure (MAOP) of > 15 bar_g.
- Leaks are counted only for the pipeline segments that are outside the fences of the gas installations. Leaks from pipeline segments inside the facility fence are not considered part of this database since it focuses on cross-country pipelines.

The database collects LOC incidents from 17 European gas companies listed below, which makes this database one of the most comprehensive of its kind. The ninth report of EGIG [1] shows the release frequency of gas pipelines and how it changed over the years since 1970. This is summarized in Table 1, which shows that generally, gas pipelines in Europe can fail at a rate in the order of 3 leaks/year/10,000 km. Recent statistics (in the last 20 years) show that the rate is a little higher than half of the average rate for the total period of > 40 years. This indicates that there is generally improvement in maintaining the integrity of the pipelines with time. Current practices could also be more robust than older ones.

The ninth report of EGIG provides also information on the distribution of the release size and cause. The distribution of the frequency by release cause and size is shown in Figs. 1 and 2 for the data in the period 1974–2013. Details of how each cause failure frequency is distributed by size are given in Table 2.

TABLE 1 EGIG Leak Frequency Since 1970

Period	Interval	No. of Incidents	Exposure[a] km/year	Leaks per km/year
1970–2007	Seventh report (38 years)	1173	3.15E+06	3.72E−04
1970–2010	Eighth report (41 years)	1249	3.55E+06	3.52E−04
1970–2013	Ninth report (44 years)	1309	3.98E+06	3.29E−04
1974–2013	40 years	1179	3.84E+06	3.10E−04
1984–2013	30 years	805	3.24E+06	2.48E−04
1994–2013	20 years	426	2.40E+06	1.78E−04

[a]*Defined as the total length of the pipeline network times the number of years the pipelines have been in operation.*

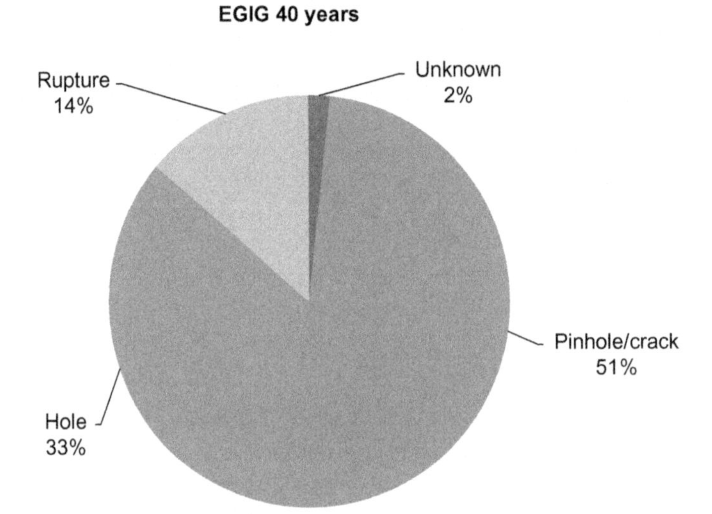

FIG. 1 EGIG release size distribution (1974–2013).

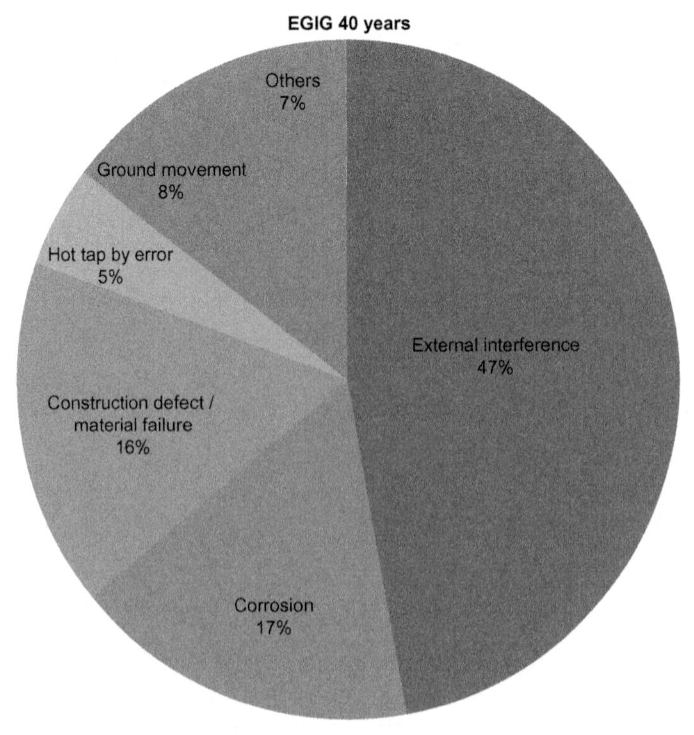

FIG. 2 EGIG release cause distribution (1974–2013).

TABLE 2 EGIG Leak Frequency for 1974–2013 by Release Cause and Size

	Failure Frequency (Leak/km/year)					
Leak Size	**External Interference**	**Corrosion**	**Construction Defect/ Material Failure**	**Hot Tap by Error**	**Ground Movement**	**Others**
Unknown	9.30E−07	9.30E−07	9.30E−07	0.00E+00	1.90E−06	4.60E−07
Pinhole/crack	4.00E−05	4.80E−05	3.60E−05	9.30E−06	5.60E−06	1.90E−05
Hole	7.60E−05	1.90E−06	1.00E−05	5.60E−06	7.40E−06	9.30E−07
Rupture	2.80E−05	4.60E−07	3.70E−06	0.00E+00	9.30E−06	4.60E−07
Total	1.40E−04	5.20E−05	5.10E−05	1.50E−05	2.40E−05	2.00E−05

The report also provides a distribution of leaks by release size for different causes [1].

The database defines release sizes as follows:

- Pinhole/crack $\leq$ 2 cm
- 2 < hole < pipe diameter
- Rupture $\geq$ pipe diameter

CONCAWE Database: 1971–2013

The CONCAWE European cross-country oil pipeline database [2] includes LOC events from pipelines that fit the following criteria:

- Onshore pipeline used for transporting crude oil or petroleum products.
- Has a minimum length of 2 km in the public domain and runs cross-country.
- Pump stations, intermediate aboveground installations, and intermediate storage facilities are included in the database as well. However, origin and destination terminal facilities and tank farms are excluded.
- The database only records spills above 1 m^3 (around 6 bbl), unless exceptional safety or environmental consequences are reported.

Based on the description above, it is clear that the database does not cover sour/wet pipelines. As such, it may not be adequate to use this database for upstream pipelines carrying sour and wet oil (that can have high corrosive rates).

CONCAWE's report no. 4/15 [2] provides a good description of the database. Fig. 3 summarizes the overall leak frequency for the period of 1971–2013. Generally, the release rate (frequency) has declined with time indicating better practices in pipeline integrity management over time.

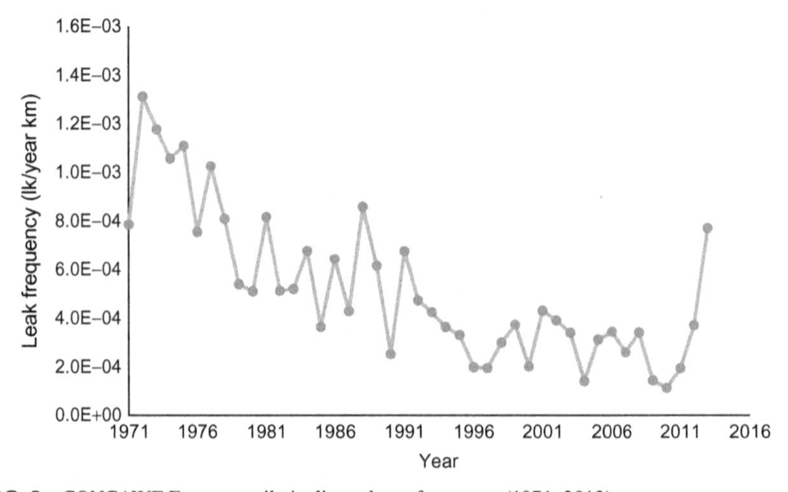

FIG. 3 CONCAWE European oil pipeline release frequency (1971–2013).

Overall, in the period indicated above, there have been a total of 524 incidents reported with a total exposure of 1.14 million km/year of pipeline network. The average release rate would therefore be around 4.6 leaks/year/10,000 km of pipeline (4.6E − 4 leaks/year/km). This is around 30% more than the general gas leak frequency based on EGIG as shown above, but it is still in the same order of magnitude. Distribution of the release rate by size and cause is shown in Table 3. The data show that the third-party external impact tends to cause large releases, while releases caused by corrosion and mechanical defect tend to be small to medium. Other causes tend to be randomly distributed.

CONCAWE database also contains information on three subcategories of pipelines. These are stabilized crude oil, white products (light clean hydrocarbon products such as gasoline and naphtha), and hot products (such as fuel oil).

The release frequency cause and size distribution of release for each category is summarized in the CONCAWE report no. 4/15 [2]. Assessment of important parameters affecting the rate of failure is also presented. It is interesting to see that the data show that the majority of releases occur at the main body of the pipelines.

UKOPA Database: 1962–2013

This database contains release statistics for onshore pipeline network in the United Kingdom [3]. Most of the pipelines in the network are for gas distribution, as shown in Table 4. However, the database also contains release incidents from other service pipelines. It is intended for land use permit processing purposes and is also used for pipeline risk assessments.

The release frequency over the years is shown in Table 5. Generally speaking, the 5-year average release frequency drops with time, but on average, the overall frequency is consistent with the CONCAWE and EGIG statistics. It is around 2.2E − 4 release/year/km (> 2 releases/10,000 km/year). The last 20-year average data are also shown in the table.

The declining rate of release shown here is a reflection of the general improvement in pipeline integrity management practices seen in the industry in recent years. It also indicates that the practices in pipeline design, manufacturing, operation, and maintenance are consistent across different places in Europe and the United Kingdom.

The release frequency distributions by release size and cause are shown in Figs 4 and 5. Most releases are small. Only 21% of total releases are medium to large release as shown in these figures. Mechanical failure, third-party external damage, and corrosion are responsible for ¾ of the total incidents reported in this database.

Canadian Release Data (BC Oil and Gas Commission 2009–12)

The BC Oil and Gas Commission 2012 Pipeline Performance and Activity Report summarizes the leak frequency data for pipeline releases in the period

TABLE 3 CONCAWE Release Frequency Data

	Leak Dimensions	Hole	Pinhole	Fissure	Hole	Split	Rupture		
Hole Type	**Equiv. Dia (in.)**	**<0.5**	**<0.5**	**0.5**	**1.7**	**7**	**FBR**	**Overall**	
Events	Number of leaks	14	33	48	104	52	59	310	100%
	% of total leaks	5%	11%	15%	34%	17%	19%	100%	
Release caused by	Mechanical	9	4	14	13	17	7	64	21%
	Operational	2	0	1	2	3	4	12	4%
	Corrosion	0	23	11	24	17	5	80	26%
	Natural hazard	0	1	2	0	2	2	7	2%
	Third party	3	5	20	65	13	41	147	47%

TABLE 4 UKOPA Database Network Size by Type of Service

Material		Pipeline Length	
		Total Length (km)	% of Total Network
Natural gas (dry)		20,388	92%
Ethylene		1141	5%
Natural gas liquids		251	1%
Crude oil (spiked)		224	1%
Others	Ethane	38	<1%
	Propylene	38	
	Condensate	24	
	Butane	20	
	Propane	20	
	Hydrogen	14	
Decommissioned		855	
Total		21,303	100%

TABLE 5 UKOPA 5-Year Average Release Frequency With Time

Period	Number of Incidents	Total Exposure (km/year)	Frequency (Leak/km/year)
1964–68	12	20,742	5.80E−04
1969–73	30	54,654	5.50E−04
1974–78	23	71,385	3.20E−04
1979–83	25	84,055	3.00E−04
1984–88	44	91,353	4.80E−04
1989–93	20	96,424	2.10E−04
1994–98	9	101,971	8.80E−05
1999–2003	5	105,808	4.70E−05
2004–08	7	107,996	6.50E−05
2009–13	12	114,480	1.00E−04
Total	187	848,868	2.20E−04
Last 20-year total	33	430,255	7.70E−05

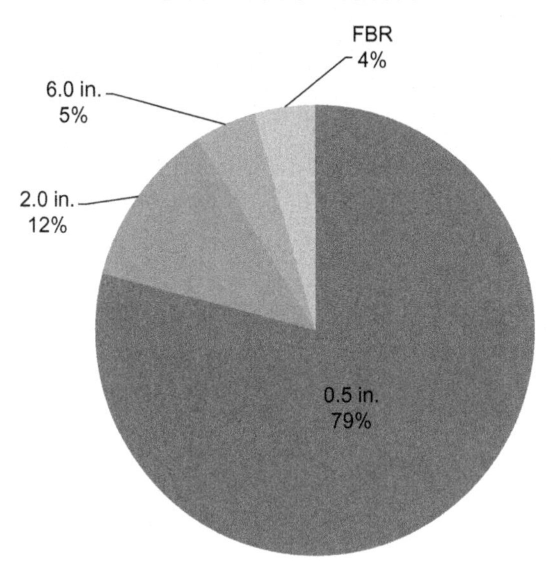

FIG. 4 UKOPA release frequency by size.

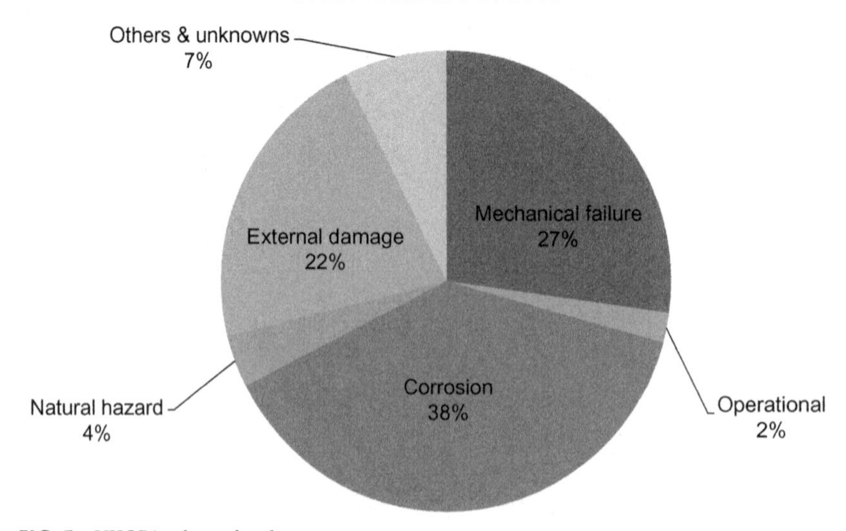

FIG. 5 UKOPA release data by cause.

TABLE 6 BC Oil and Gas Commission Release Frequency

	Year				
Incident Cause	**2012**	**2011**	**2010**	**2009**	**Total**
Number of incidents	27	34	50	37	148
Total pipeline length (km/year)	40,125	39,023	36,361	35,868	151,377
Release frequency (leak/year)	6.7E−04	8.7E−04	1.4E−03	1.0E−03	9.8E−04

of 2009–12 [4]. Table 6 summarizes the total release frequency for each year. It is clear that this database shows slightly higher than other previous rates. However, it is in the same order of magnitude.

The reported incidents include leaks from hydrocarbon services (oil and gas) and water. The majority of the network is for natural gas (sour and sweet) as shown in Fig. 6. The distribution of the total release by cause of incidents for 4 years reported above is given in Fig. 7. Data show that natural gas tends to have lower frequency of release compared with other types of services, as shown in Fig. 8, which are consistent with the previous databases. Sour-gas pipelines are twice more likely to leak than sweet-gas pipelines, while crude pipelines are even an order of magnitude higher. This might be attributed to the higher corrosion rate associated with wet/sour crude and gas services.

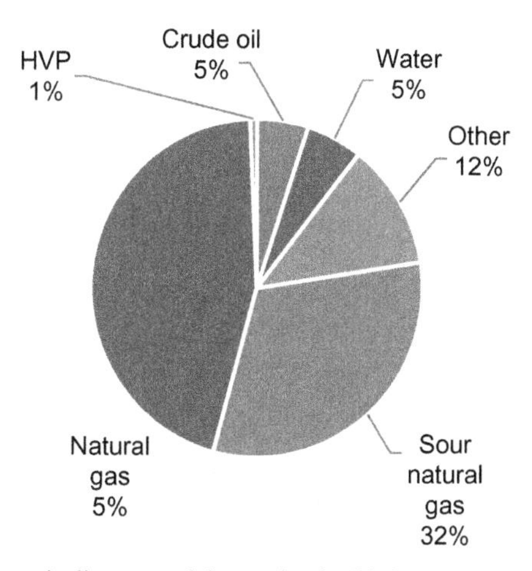

FIG. 6 BCE oil and gas pipeline network by service for 2012.

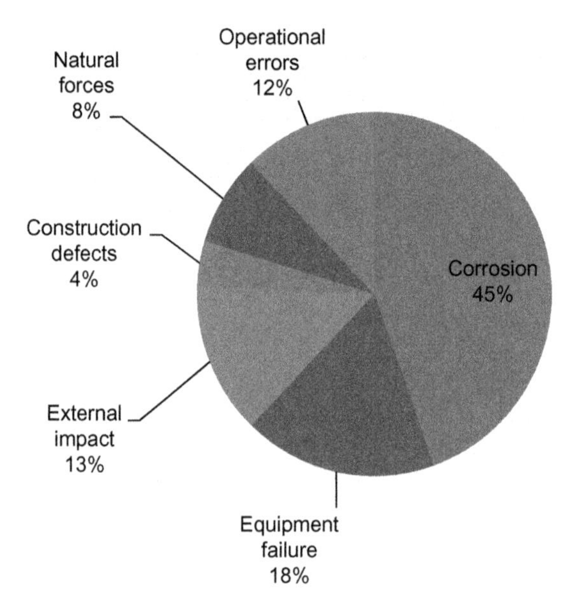

FIG. 7 BCE oil and gas release data by cause.

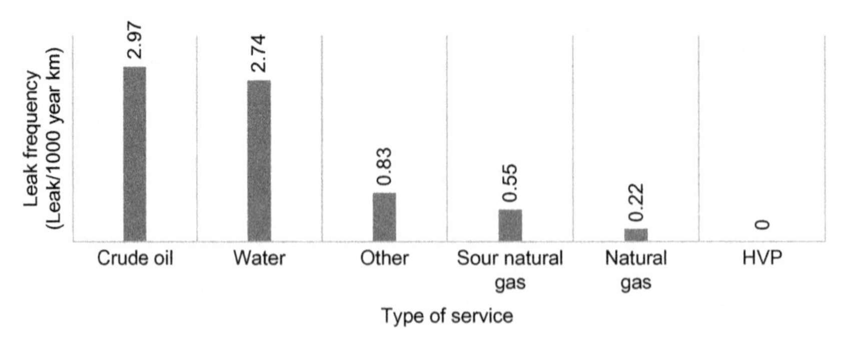

FIG. 8 Release frequency by type of service for 2012 data of the BC database. HVP refers to high-vapor-pressure material (i.e., volatile material).

USA-DOT PHMSA, 1995–2014

The PHMSA of the US Department of Transportation (DOT) collects pipeline-related incidents for the massive 2.6 million miles of pipelines carrying hazardous materials. PHMSA was created in 2004, and it covers all kinds of pipelines operating in the United States including the following:

- Gas: Sour and sweet gas, transmission, collection, and distribution network
- Liquid: Sour crude, stabilized crude, petroleum/refined products, NGL, and CO_2
- Onshore and offshore sections

The available information online from the database website shows a total of 10,848 incidents, for around 80 million km/year of pipeline exposure, and covers 20 years of operation [5]. However, the information available online does not contain data on release size, but spill volume is available, and failure causes are quite detailed. The overall release frequency was calculated based on the available data and summarized in Table 7.

PHMSA defines different types of services as follows:

- Crude oil is liquid petroleum out of the ground, which includes sour and wet crude. Refined oils manufactured from crude oil are included in refined PP.
- Refined PP is petroleum products that are liquid at ambient conditions, such as gasoline, diesel, jet fuel, kerosene, and fuel oil. These are clean and sweet materials.
- HVL flammable/toxic includes highly volatile liquids (HVL), flammable, and toxic liquids, such as propane, ethane, butylene, and anhydrous ammonia.
- CO_2: this is transferred in the liquid state.

Table 7 shows that crude oil has a failure frequency higher than any other services, and it is consistent with the crude failure frequency of the Canadian databases shown above. Refined products have high frequency than the oil pipelines in CONCAWE. The reason could be attributed to the reporting criteria. The CONCAWE reports only incidents with spill size >6 bbl of oil, while the PHMSA reports all incidents including the ones causing negligible spill sizes. A comparison was conducted between different databases but for normalized data (same incident categories and reporting criteria) and found that the frequencies of PHMSA, CONCAWE, and pipeline leak data collected by the Canadian

TABLE 7 US DOT Pipeline Leak Frequency From PHMSA Database

Category	Available Year	No. of Incidents	PL Exposure (km/year)	Frequency (Leak/km/year)
Crude oil (sour)	10 years	1914	851,000	2.25E−03
Refined products	10 years	1481	1,010,000	1.46E−03
HVL flammable-toxic	10 years	566	907,000	6.24E−04
Total hazardous liquids	10 years	4057	2,840,000	1.43E−03
Gas transmission	10 years	2182	10,400,000	2.10E−04
Gas gathering and distribution	20 years	2573	62,600,000	4.11E−05
Total	20 years	7009	39,300,000	1.78E−04

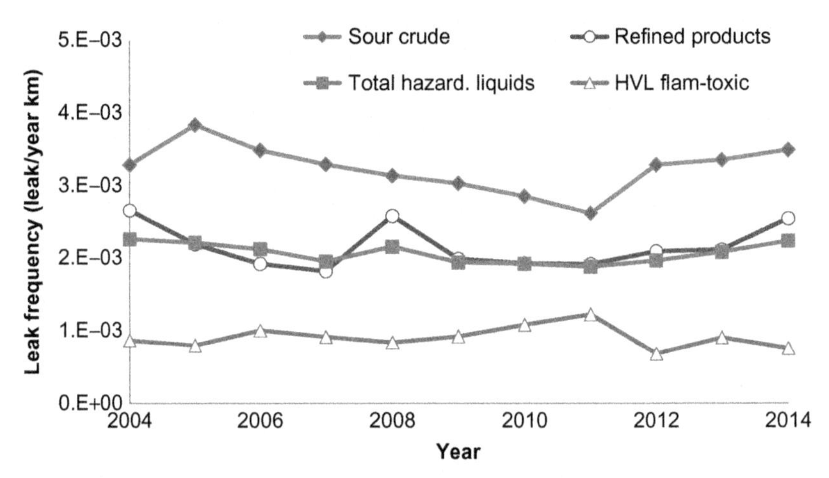

FIG. 9 Annual leak frequency for liquid pipelines in PHMSA database.

National Energy Board are consistent and very similar [6]. The PHMSA data show that the gas transmission failure rates are similar to the European EGIG database. Gas distribution network has much lower failure rates.

Over the reporting period, for which data are available, it seems that the release frequency remained steady for the hazardous liquid pipelines as shown in Fig. 9. The same applies for gas distribution, as shown in Fig. 10. As for gas gathering and transmission, the trend seems to be increasing in recent years (Fig. 10).

PHMSA online database does not show information on release size, but using the spill size information, an equivalent release size distribution was defined. This is illustrated in Fig. 11, which shows the distribution of spill size for crude oil data. For the purpose of estimating release size distribution, it was assumed that

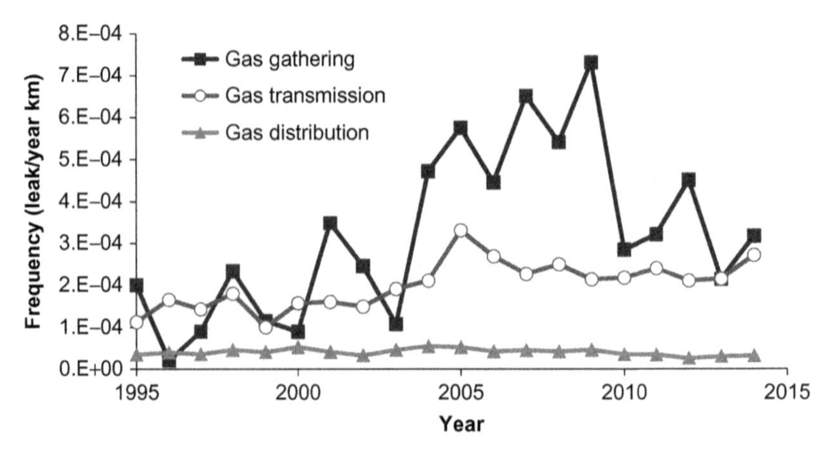

FIG. 10 Annual leak frequency for gas pipelines in PHMSA database.

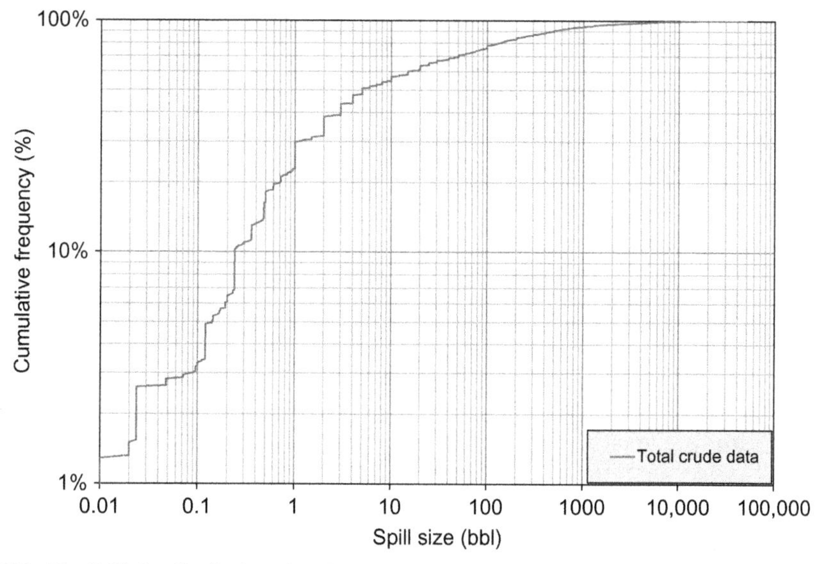

FIG. 11 Spill size distribution of crude release from PHMSA database.

- small leaks are all leaks below 10 bbl of spill size and that is equivalent to ½ in release size,
- medium releases are all releases between 10 and 100 bbl of spill size and this is equivalent to 2 in. of release size,
- large releases are all releases with spill size between 100 and 1000 bbl and that is equivalent to 6 in. release size,
- full bore ruptures are for all releases larger than 1000 bbl.

These are approximations but are very useful in the absence of data on the release size distribution. Using this approach, the crude release size distribution is estimated as shown in Table 8. Similar assessments were performed for the rest of the pipeline categories based on data shown in Fig. 12. The results are summarized in Fig. 13. For gas releases, the distribution of EGIG was assumed due to the lack of data.

TABLE 8 Release Size Distribution for Crude Releases of PHMSA Database

Leak Category	Release Size (in.)	Spill Size (bbl)	Frequency %
Small	½	10	50%
Medium	2.0	100	25%
Large	6.0	1000	20%
FBR	FBR	1000+	5%

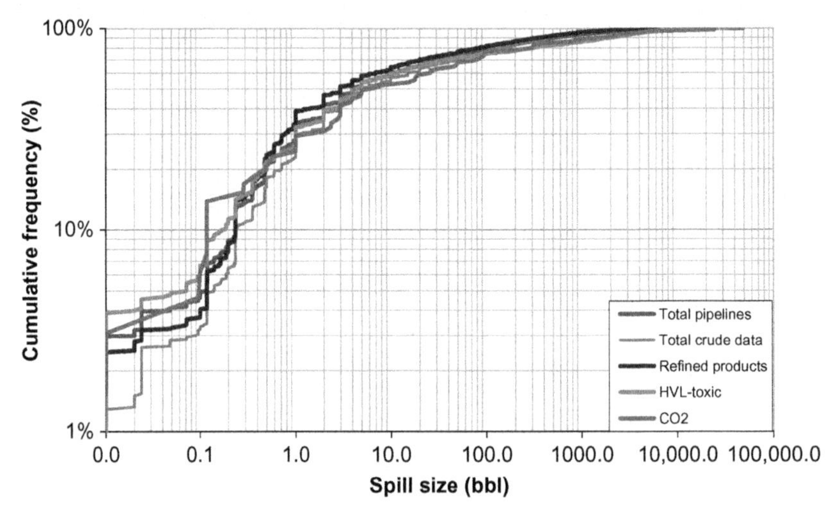

FIG. 12 Spill size distribution of different hazardous liquids from PHMSA databases.

Causes of releases for different pipeline categories in PHMSA are given in Figs. 14–20. The database divided external impact effects into excavation damage effects and other outside force effects. Original data even divide each one of these main categories into several subfactors. For example, corrosion is divided into internal, external, and stress corrosion cracking.

All release frequency databases analyzed in this assessment are summarized in Figs. 21 and 22. The comparison shows the following:

- Generally, sour/wet pipelines have higher frequency of release compared with clean-service pipelines.
- Liquid pipelines tend to have higher release frequency than gas pipelines.
- US and Canadian pipeline release frequencies are similar and tend to be higher than the European pipeline failure rates.
- Recent failure rates tend to be lower than older values, even for the same database and pipeline categories/services.
- The UKOPA data are generally lower than the other databases but tend to be similar to the PHMSA gas frequencies, and that could be due to the fact that it is mainly abased on gas transmission and distribution network.

LEAK DETECTION MECHANISMS

Leak detection mechanisms were reported in some of the databases. Table 9 shows the main mechanism for detecting pipeline leaks that were reported. For the purpose of the analysis presented here, some modifications/new definitions were introduced to combine the data and present it in a consistent manner.

Average (combined) data are shown in Fig. 23. The data presented in the table above and the figure below show the following:

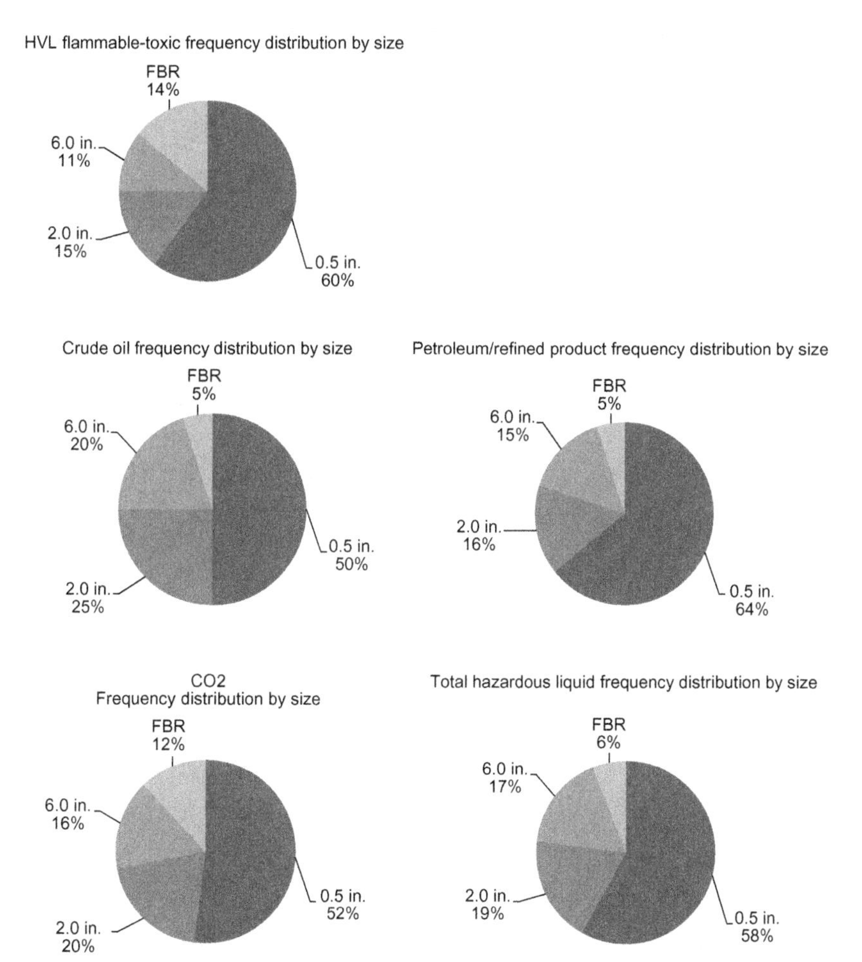

FIG. 13 Release size distribution for PHMSA database.

- External party (third party and public) and company personnel are the main detection mechanisms for pipeline leaks, followed by operational mechanisms such as pressure testing, SCADA, and cathodic protection systems.
- In-line inspection (ILI) has also detected leaks but at a very low percentage.
- The Canadian database and CONCAWE have explicitly reported data on pipeline leak detection systems (LDS), and it shows that only 1% and around 13% of total leaks have been picked by LDS in both databases, respectively. The big difference between both cases could be attributed to different factors, including that not all pipelines are equipped with LDS in the case of the Canadian data compared with the CONCAWE data. CONCAWE reports indicate that pipeline LDS are becoming more effective [7].

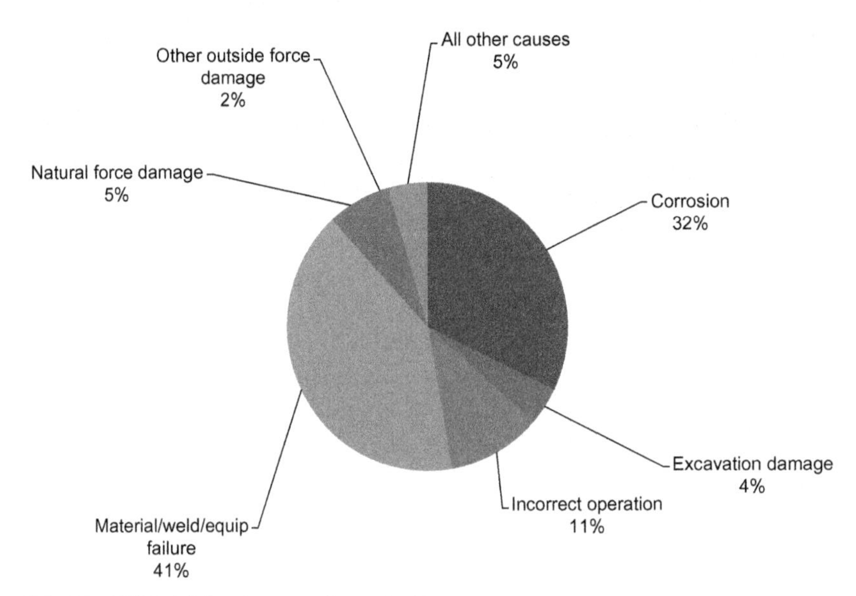

FIG. 14 PHMSA failure by cause for crude oil.

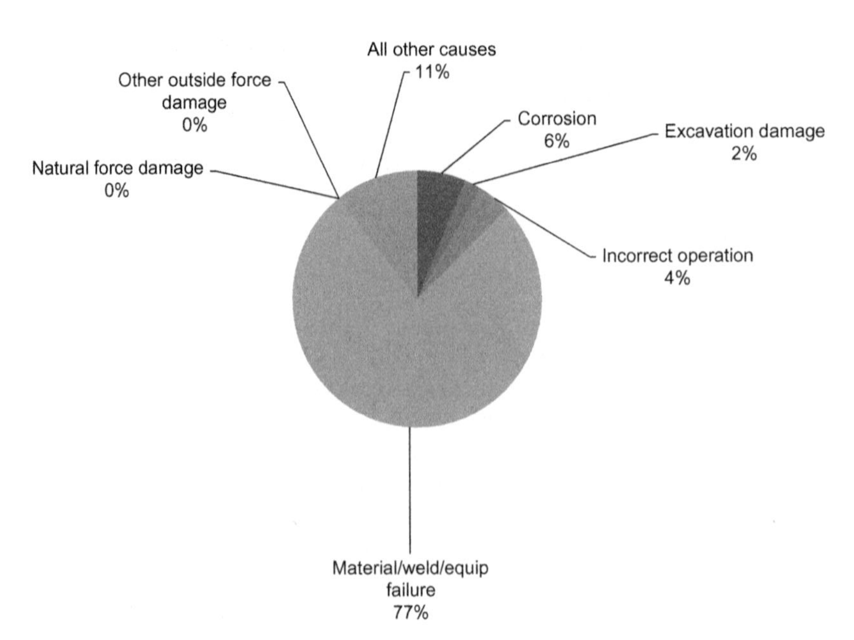

FIG. 15 PHMSA failure by cause for CO_2.

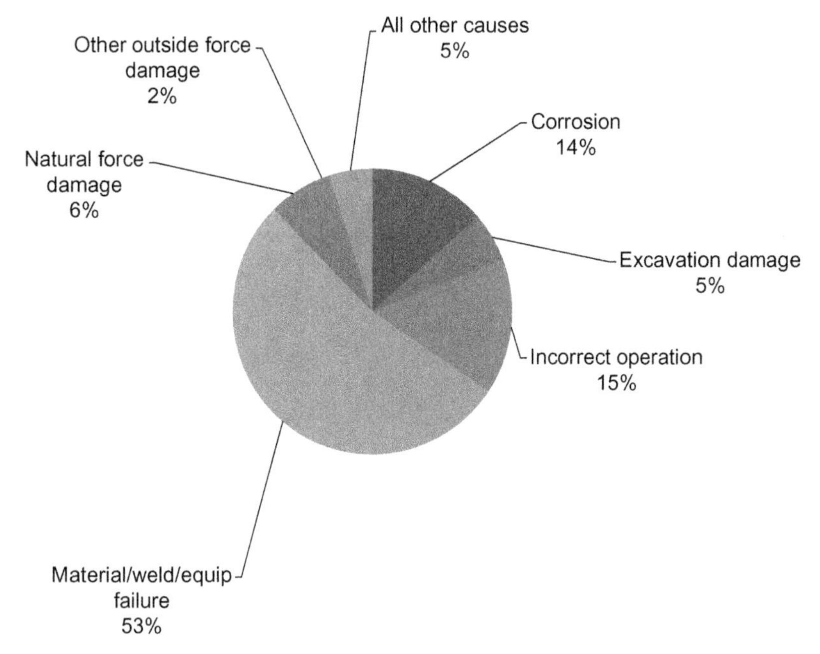

FIG. 16 PHMSA failure by cause for petroleum products.

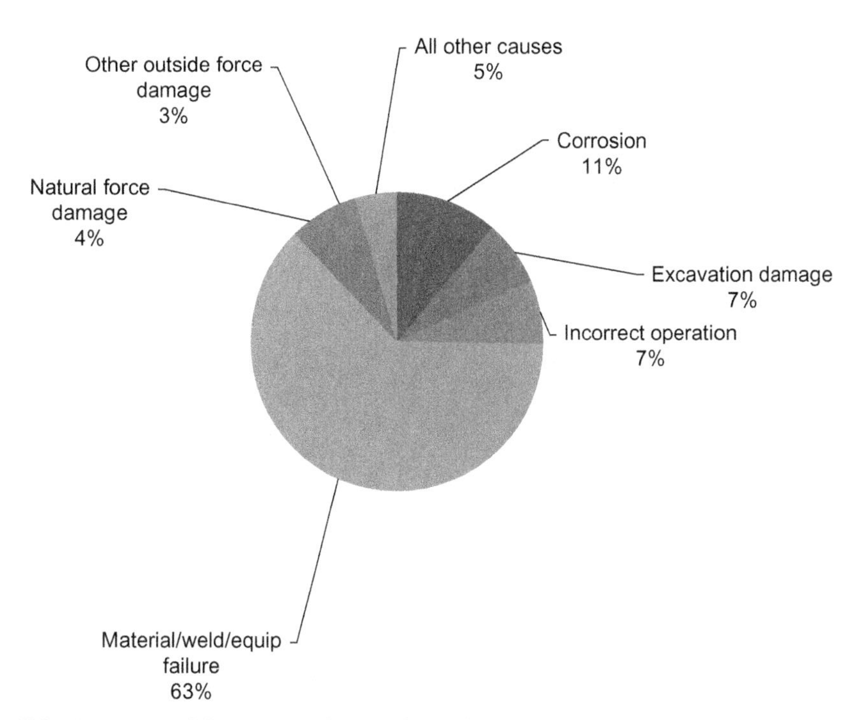

FIG. 17 PHMSA failure by cause for HVL flammables and toxics.

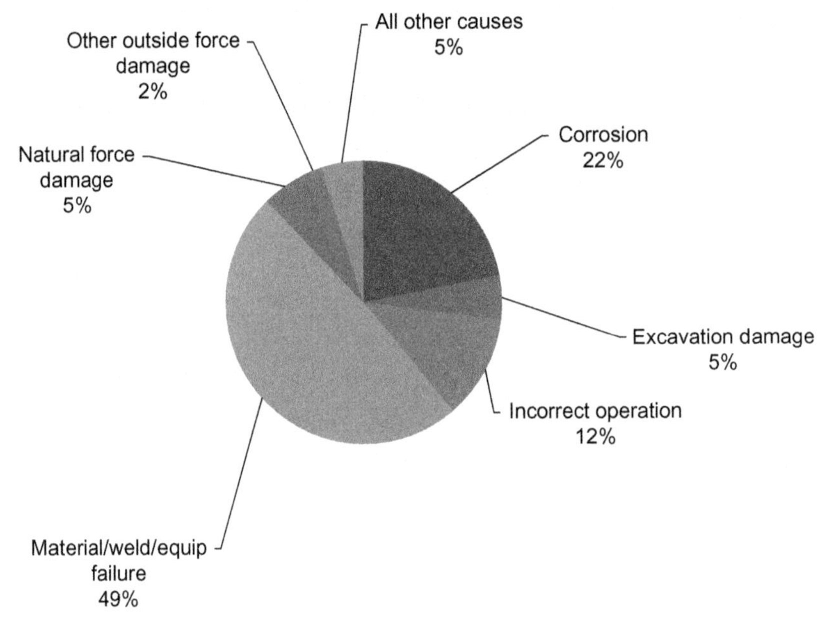

FIG. 18 PHMSA failure by cause for total hazardous liquid pipelines.

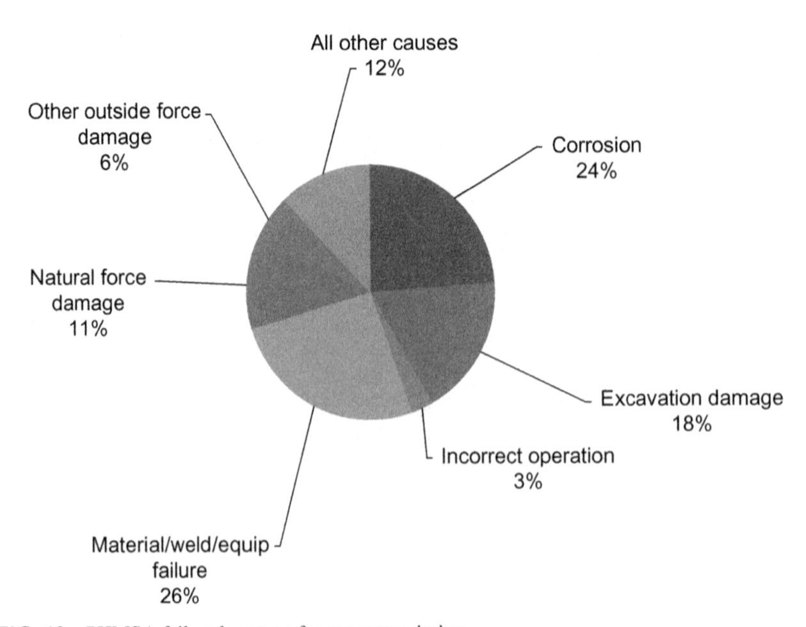

FIG. 19 PHMSA failure by cause for gas transmission.

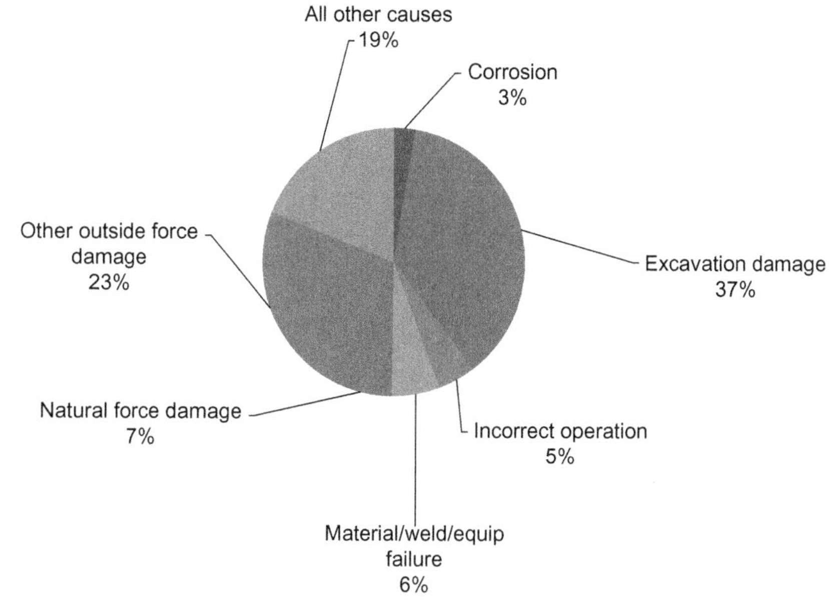

FIG. 20 PHMSA failure by cause for gas gathering and distribution.

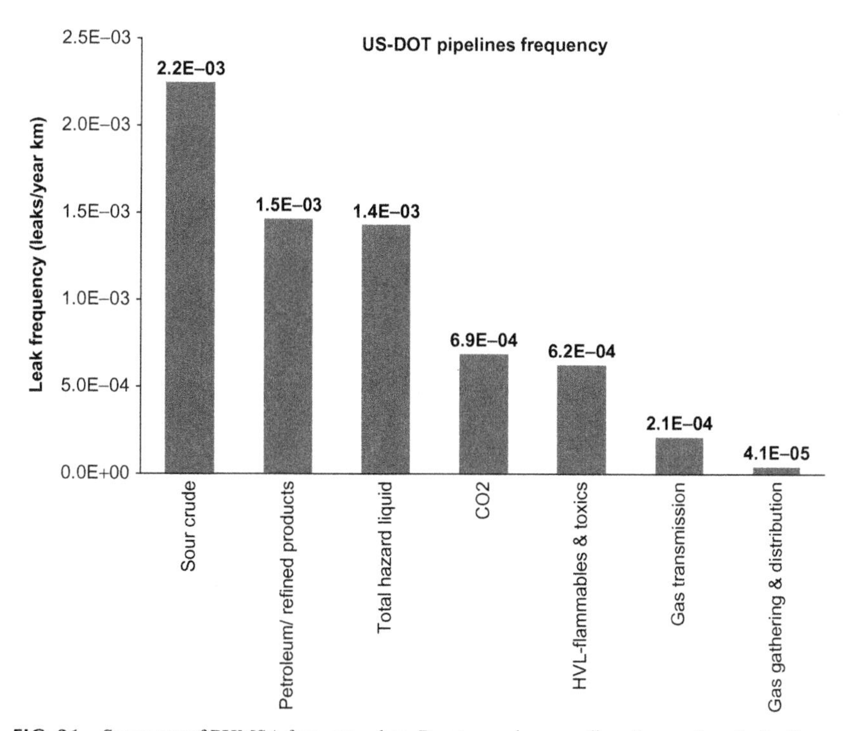

FIG. 21 Summary of PHMSA frequency data. Due to number rounding, the numbers in the figure are slightly different from the ones in the table. Table numbers are the reference.

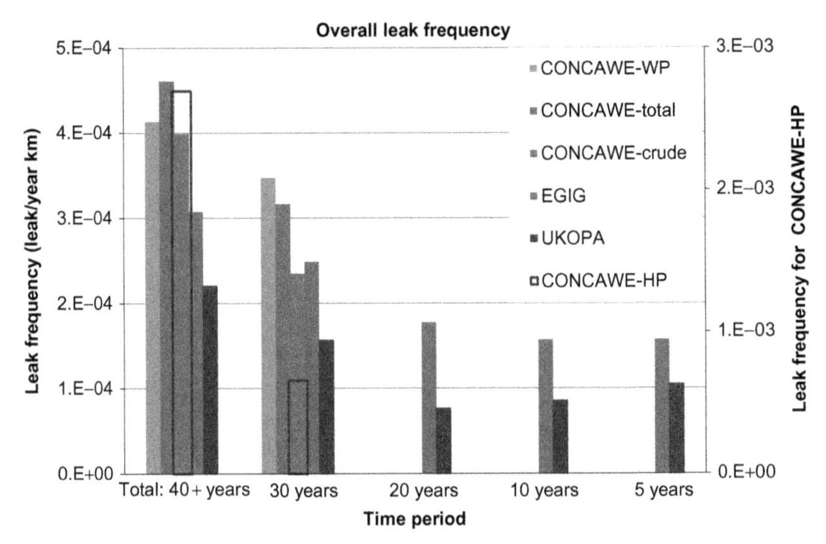

FIG. 22 Summary of release frequency databases.

- CONCAWE recent reports indicate that LDS are becoming more effective in the detecting leaks for liquid pipelines, based on data reported recently in the last 5 years [2,7].

However, it is still interesting to see that engineering systems did not contribute much to leak detection regardless of the cause, and this should be considered when designing pipelines and assessing their risks. Description of leak detection systems is given in the literature [8,9].

IGNITION PROBABILITIES

Databases have reported statistics on ignition. The most comprehensive of which is the EGIG data [1]. For the purpose of this report, the EGIG ignition data have been analyzed and presented in the format shown in Table 10. The information in the table is based on EGIG data published in EGIG report with some analysis and assumptions performed by the author of this book.

To be conservative in risk calculations, toxic releases have been assumed to have lower ignition probabilities, while flammable-only releases were assumed to have the higher end of the ignition probability range. This is conservative as it will predict higher risk levels.

The overall ignition probability of EGIG data is around 5.0%, which is consistent with the UKOPA data (4.7%), but much higher than the total ignition probability of liquid releases reported in CONCAWE (~1%) [1–3]. This is used to develop a model for ignition probabilities that depend on pipeline-specific conditions (Fig. 3). For liquid releases, the model assumes the ignition probability to be 20% of the gas one for large releases, based

TABLE 9 Pipeline Leak Detection Mechanism Data

Mechanism		EGIG	UKOPA	CONCAWE				Canadian Data
				Crude	WP	HP	Total	
External party	Public, police, land owner	46%	30%	47%	44%	50%	46%	11%
Company personnel	Routine patrol and monitor, contractors and company staff	40%	26%	42%	31%	39%	35%	73%
Operational measures	Cathodic protection, pressure testing, internal inspection, and SCADA	0%	7%	3%	5%	9%	5%	14%[a]
PL LDS and automatic system		0%	0%	8%	19%	1%	13%	1%
In-line inspection (ILI)		2%	5%	0%	1%	1%	1%	0%
Unknown/others		13%	32%	0%	0%	0%	0%	0%
Total		100%	100%	100%	100%	100%	100%	99%

[a] *This is mainly related to SCADA in the case of the Canadian database.*

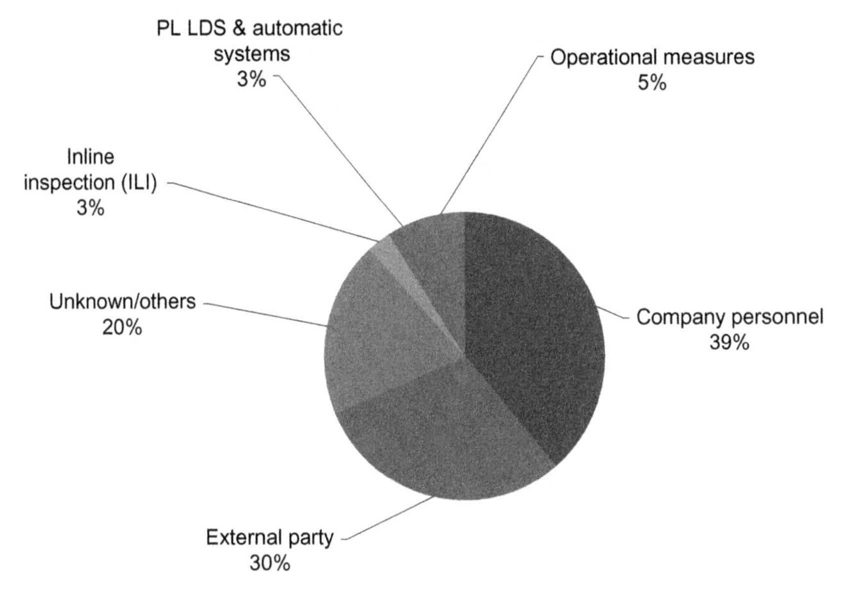

FIG. 23 Average leak detection data for combined databases.

on the comparison of the total ignition probabilities reported in EGIG and CONCAWE reports.

For small to medium liquid releases, ignition probability would be much lower than that, and for the purpose of this report, it is assumed to be 5% of the gas ignition probability. HVL (such as NGL) would still have ignition

TABLE 10 Pipeline Ignition Probabilities for Gases

Release Size	PL Size (in.)	Toxic	Flammable		Average
			>35 bar	≤35 bar	
0.5 in.		4%	4%	4%	4%
2.0 in.		2%	2%	2%	2%
6.0 in.		10%	40%	10%	20%
Full bore rupture	<8 in.	10%	40%	10%	20%
	8–12 in.	10%	40%	10%	20%
	12–24 in	14%	48%	12%	24%
	24–30 in.	32%	55%	43%	43%
	>30 in.	32%	55%	43%	43%
	Unknown/others	14%	48%	12%	24%

probabilities similar to that of gas, since these materials have a tendency to disperse over a larger area, which increases their chance of getting ignited (Fig. 24).

DATABASES ANALYSIS

Different databases have reported information on factors contributing to pipeline failures. The data reported were analyzed, and for the purpose of this assessment, a frequency factor (FF) was defined as follows:

$$\text{Frequency factor} = \frac{\text{Failure frequence of data segment}}{\text{Failure frequency of total datbase or pipeline category}}$$

For example, the total pipeline leaks reported in EGIG due to external third-party interference are around 1.6E−4 leaks/km/year, but if we consider pipelines with size range from 11 to 17 in., the frequency will be 1.2E−4 leaks/km/year. Therefore, the factor would be as given below (Table 11):

$$\text{Frequency factor} = \frac{1.2\text{E}-4}{1.6\text{E}-4} = 0.77$$

Using the data reported in EGIG, UKOPA, and CONCAWE, frequency factors (FFs) were developed as shown in external damage to gas pipelines using the data reported in EGIG. Detailed analysis was conducted for all available factors, and the results are presented in Figs. 25–36.

Assumptions, definitions, and modifications used in preparing Figs. 25–36 are summarized in Table 12 for oil and gas pipelines. Mainly, oil FFs were based on CONCAWE data, and gas FFs were based on EGIG and UKOPA.

PIPEFAIT

Using the data analysis shown in the previous sections of this chapter, a new tool called pipeline failure and integrity tool (PipeFAIT) was developed. The tool uses the base frequencies of relevant pipeline categories from the corresponding databases described above and then modifies the frequency using the frequency factors defined earlier to produce a customized pipeline leak frequency based on the conditions and operating parameters of each pipeline (or pipeline segment).

A matrix for FF is defined with relevant factors being assigned to different release causes as shown in Table 13, where the value inside the cell indicates FF value. The matrix shows the following:

- Pipeline age (in terms of year of construction) affects the corrosion and material defect causes.
- Pipeline size affects external impact and natural force causes.
- Pipeline wall thickness affects external impact, corrosion, and operational error causes.

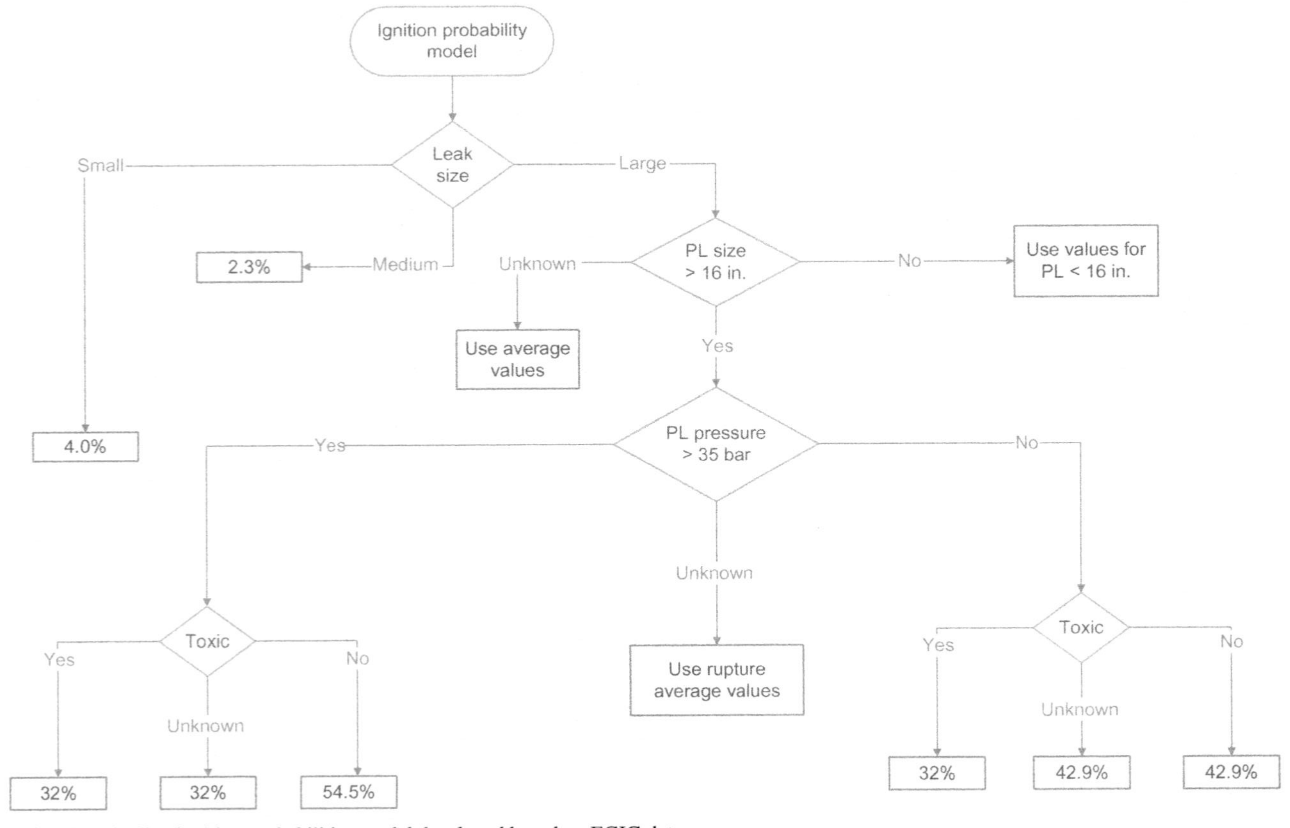

FIG. 24 Pipeline ignition probabilities model developed based on EGIG data.

TABLE 11 Frequency Factor for EGIG External Impact per PL Diameter

Diameter Category	Frequency (Leak/km/year)	Factor
<5 in.	5.0E−04	3.21
5–11 in.	2.7E−04	1.73
11–17 in.	1.2E−04	0.77
17–23 in.	5.0E−05	0.32
23–29 in.	3.0E−05	0.19
29–35 in.	1.5E−05	0.10
35–41 in.	5.0E−06	0.03
41–47 in.	0.0E+00	0.00
≥ 47 in.	0.0E+00	0.00
Total	1.6E−04	1.00

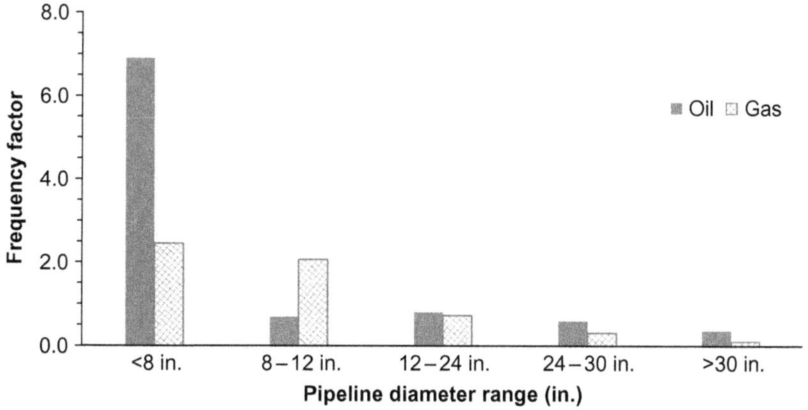

FIG. 25 FF for external damage per PL size.

- Pipeline external coating, backfill around buried pipelines, and material grade affect corrosion causes only.
- Area classification around the pipeline, burial depth for buried pipelines, and facility type affect external impact causes only.

Relevant factors are applied to different causes as shown in Table 13. The overall factor for each cause is then calculated by multiplying all factors together. For example, the overall factor for external impact is the result of the multiplication of all factors related to pipeline size, wall thickness, area classification, and

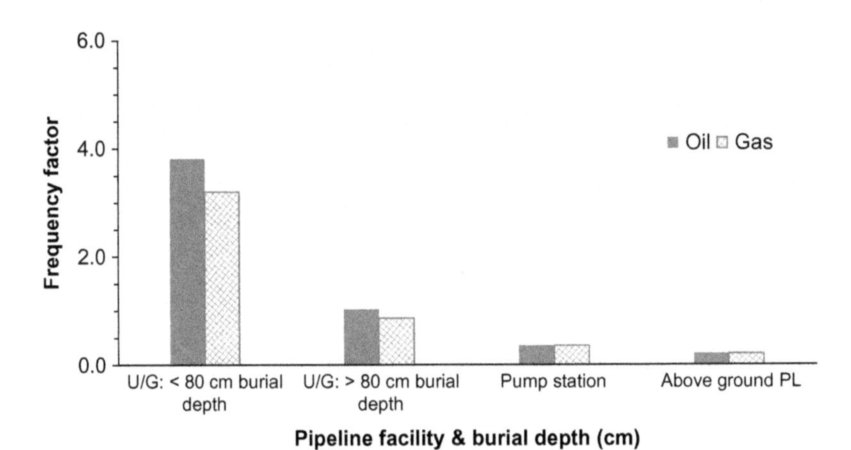

FIG. 26 FF for external damage per PL facility type and burial depth.

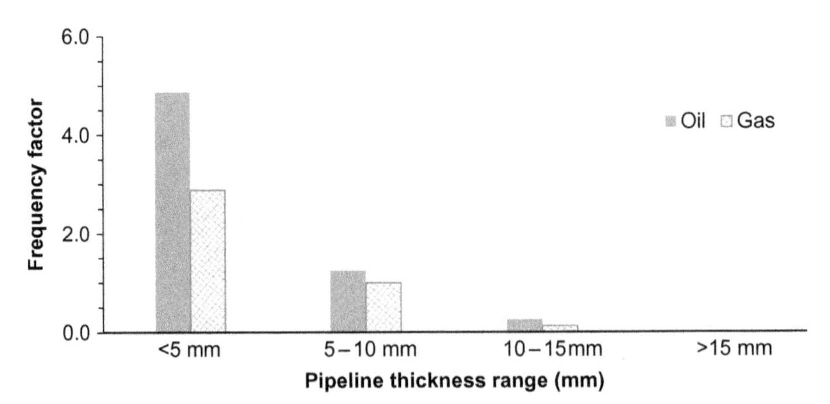

FIG. 27 FF for external damage per PL wall thickness.

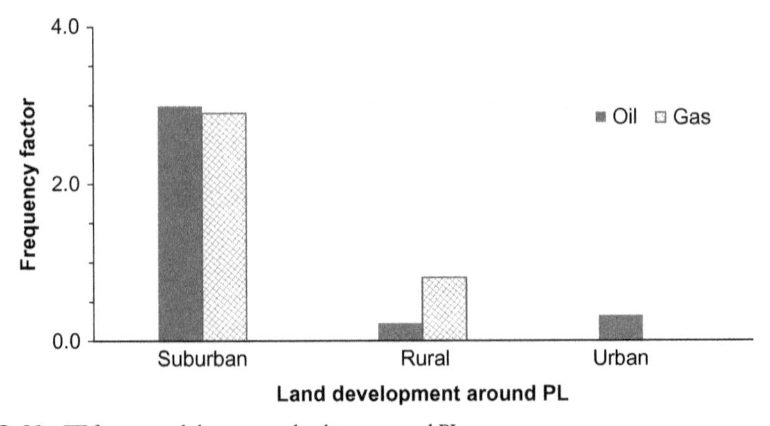

FIG. 28 FF for external damage per land type around PL.

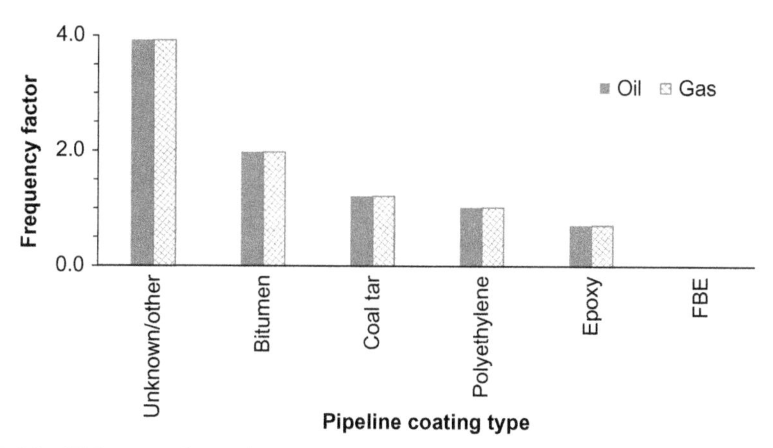

FIG. 29 FF for external corrosion per PL coating type.

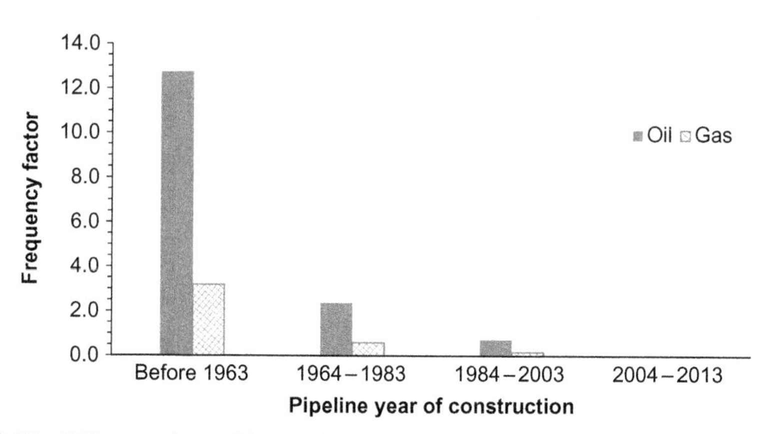

FIG. 30 FF for corrosion per PL year of construction.

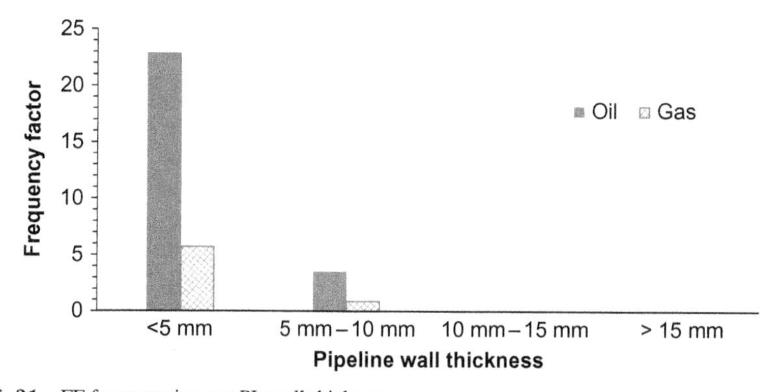

FIG. 31 FF for corrosion per PL wall thickness.

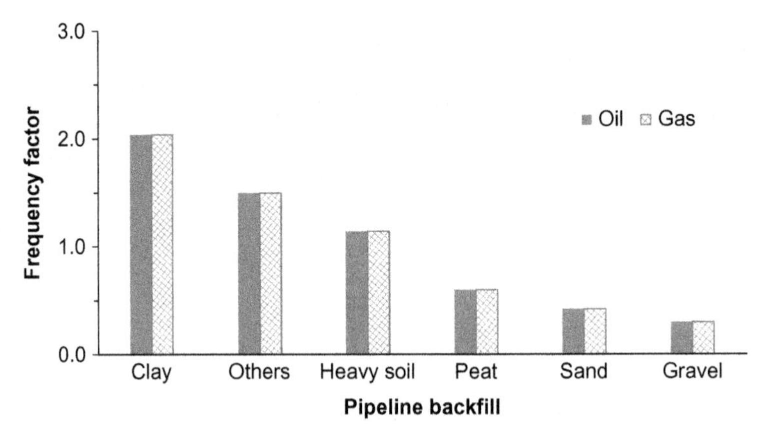

FIG. 32 FF for corrosion per PL backfill type.

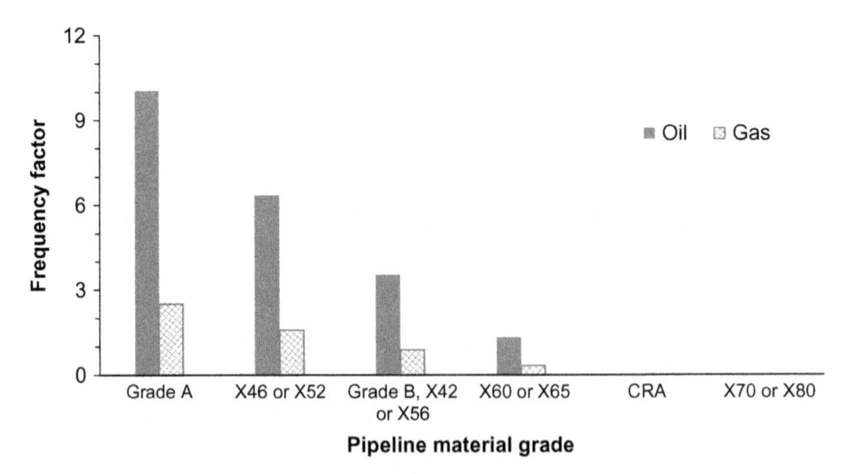

FIG. 33 FF for corrosion per PL material grade.

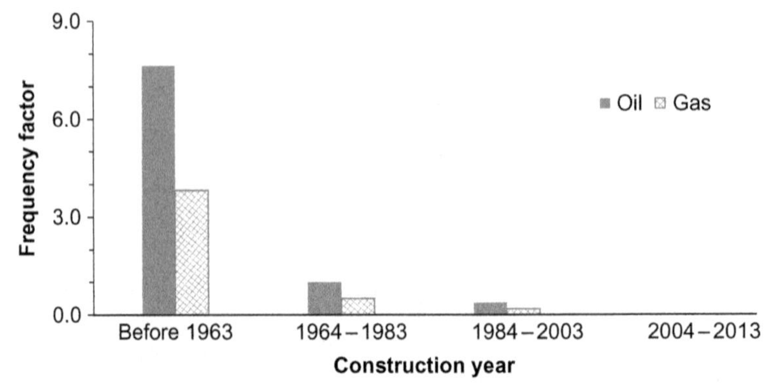

FIG. 34 FF for construction defect per PL year of construction.

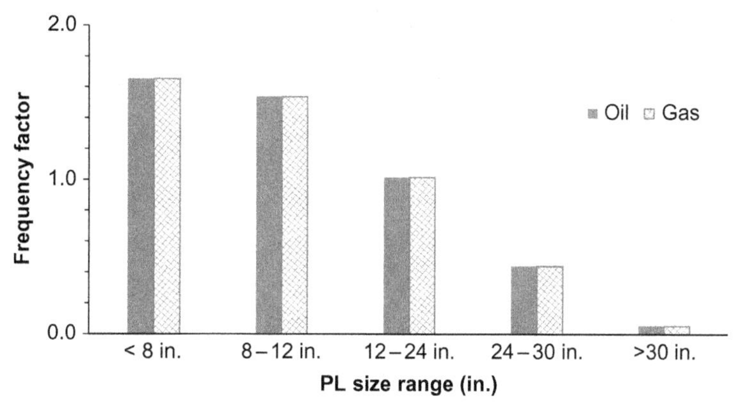

FIG. 35 FF for natural force impact per PL size.

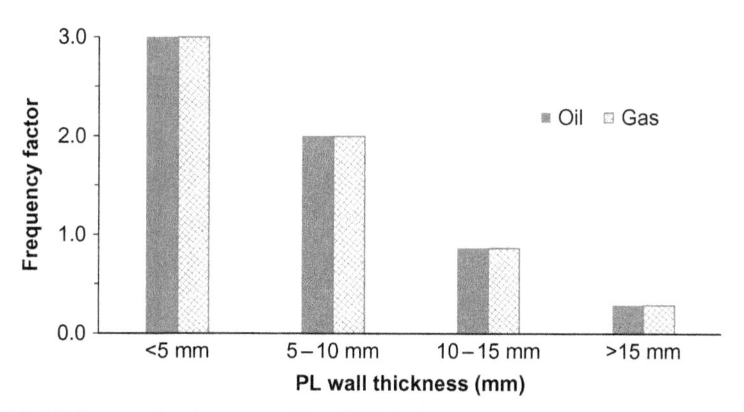

FIG. 36 FF for operational errors per PL wall thickness.

burial depth together. The result is the overall frequency factor (FFo), which is then multiplied by the frequency of the pipeline due to external impact to produce the modified frequency, and so on, for the rest of the causes.

If FFo is greater than one, then the frequency of that specific cause will be higher than the value in the original database and vice versa. In the table below, for example, the external impact has FFo of 4.61, which means that the frequency is 4.61 larger than the corresponding value in the original value based on average pipeline failure rates for that specific pipeline category. On the other hand, corrosion's FFo is 0.01, which means that the pipeline frequency of leak due to corrosion is two orders of magnitude lower than the original value.

The tool allows the user to choose values of each PL parameter, and then, it automatically reads the relevant value of FF based on the data shown in Figs. 25–36. Based on the type of service and pipeline category, a relevant frequency is chosen from the databases that were analyzed earlier in this chapter

TABLE 12 Bases and Assumptions Used in FF Assessments

Figure No.	Description	Oil FF Assumptions	Gas FF Assumptions	Notes
Fig. 25	External damage per PL size	Based on data for total liquids in CONCAWE	Based on data in EGIG	
Fig. 26	External damage per PL facility type and burial depth	Data are based on CONCAWE for facility type and EGIG for burial depth impact, with slight modifications	Data are based on CONCAWE for facility type and EGIG for burial depth impact, with slight modifications	External impact is not dependent on pipeline operating conditions, but external forces. So assumption of similarity in FF is valid
Fig. 27	External damage per PL wall thickness	Based on UKOPA data, as it contains some liquid pipelines in it	Based on EGIG data as it is mainly for gas	Difference between EGIG and UKOPA is assumed to be attributed to the presence of liquid pipelines; hence, UKOPA was used to represent liquid pipelines
Fig. 28	External damage per land type around PL	Data are based on UKOPA, with slight modifications to ensure consistency between both cases (oil and gas)	Data are based on EGIG information with slight modifications to ensure consistency between both cases (oil and gas)	The following definitions apply • Rural areas: such as agricultural, barren, forest, hills, water body • Suburban area: such as industrial or commercial, residential low-density areas • Urban areas: such as residential high density
Fig. 29	External corrosion per PL coating type	Similar to the gas data	Based on combined data of UKOPA and EGIG	This factor is related to external conditions and is therefore expected to be similar for all pipelines regardless of the pipeline service

Fig. 30	Corrosion per PL year of construction	Assumed to be four times larger than gas data due to high corrosive environments inside the pipeline	Based on EGIG data	Based on data from PHMSA, corrosion frequency of total liquid is 6.29 times the corrosion frequency of gas gathering and transmission. So, oil is more corrosive than gas. On the other hand, the ratio of corrosion of CONCAWE/EGIG is 2.2. So, the average of both the United States and European data is around 4, and this was the factor that was used in the analysis here
Fig. 31	Corrosion per PL wall thickness	Assumed to be four times larger than gas data and is applicable	Based on EGIG data	Based on data from PHMSA, corrosion frequency of total liquid is 6.29 times the corrosion frequency of gas gathering and transmission. So, oil is more corrosive than gas. On the other hand, the ratio of corrosion of CONCAWE/EGIG is 2.2. So, the average of both the United States and European data is around 4, and this was the factor that was used in the analysis here
Fig. 32	Corrosion per PL backfill type	Similar to the gas data	Based on combined data of UKOPA and EGIG	This factor is related to external conditions and is therefore expected to be similar for all pipelines regardless of the pipeline service
Fig. 33	Corrosion per PL material grade	Assumed to be four times larger than gas data due to high corrosive environments inside the pipeline	Based on EGIG data	Based on data from PHMSA, corrosion frequency of total liquid is 6.29 times the corrosion frequency of gas gathering and transmission. So, oil is more corrosive than gas. On the other hand, the ratio of corrosion of CONCAWE/EGIG is 2.2. So, the average of both the United States and European data is around 4, and this was the factor that was used in the analysis here

(Continued)

TABLE 12 Bases and Assumptions Used in FF Assessments—cont'd

Figure No.	Description	Oil FF Assumptions	Gas FF Assumptions	Notes
Fig. 34	Construction defect per PL year of construction	Assumed to be twice more likely to fail compared with gas pipelines	Based on EGIG data	Comparing CONCAWE to EGIG, oil PL is twice more likely to fail due to construction defects
Fig. 35	Natural force impact per PL size	Similar to the gas data	Based on combined data of EGIG	This factor is related to external conditions that depend mainly on the pipeline size and is therefore expected to be similar for all pipelines regardless of the pipeline service
Fig. 36	Operational errors per PL wall thickness	Similar to the gas data	Based on combined data of EGIG	This factor is related to external conditions that depend mainly on the pipeline size and is therefore expected to be similar for all pipelines regardless of the pipeline service

TABLE 13 FF for Different Leak Causes as Applied in PipeFAIT

PL Parameter	Value					
Pipeline Category	**Gas**					
Pipeline Service	**Gas Transmission**	**External Impact**	**Corrosion**	**Material/ Construction Defect**	**Incorrect Operations**	**Natural Forces**
Pipeline age	30–10 years		0.18	0.18		
Pipeline size	12–24 in.	0.73				1.02
Wall thickness	<5 mm	2.88	1.31		3	
External coating	FBE		0.07			
Area classification	Suburban	2.9				
Facility type	Underground pipe	0.87				
Backfill	Peat		0.63			
Burial depth	>80 cm burial depth	0.87				
Material grade	Grade A		1.10			
Total factor		4.61	0.01	0.18	3	1.02

and is presented in a matrix format showing the contribution of each cause versus release size distribution as shown in Table 14.

The tool will then apply relevant FFo to each cause and adjust the frequencies accordingly to produce the modified pipeline leak frequency matrix in terms of release size and release cause distribution as shown in Table 15.

The leak frequency matrix presents frequencies in a format that is ready for use in risk assessment, in terms of total frequency and release size distribution. It also defines the contribution of each cause, which helps risk analysts to define the right mitigation measures to target the most relevant/highest contributing causes of pipeline failures. In doing so, PipeFAIT provides very useful information and defines the potential modes of failures for the pipeline and its expected integrity level/status through this detailed assessment of its potential failure frequencies.

PipeFAIT defines three pipeline categories for gas, oil, and liquid. It also defines the following pipeline service, for which a dedicated original frequency matrix is available in the tool, based on available databases above:

- Sweet gas
- Gas transmission
- Gas gathering
- CO_2
- Sweet/stabilized crude
- Sour crude
- Petroleum and refined products
- HVL flammables and toxics (such as NGL and ammonia)
- White product (such as naphtha and gasoline)
- Hot product (such as heavy gas oil and fuel oil)

The tool also defines two pipeline pressure categories ($>$ 35 bar_g and $<$35 bar_g) for ignition probability calculations. Given all the parameters and their variations available in the tool, a total of $>$9,000,000 (nine million) pipeline combinations are produced by the tool. This translates to unlimited potential of the tool, and no different pipelines will have similar frequencies. For ignition probabilities, $>$60 different combinations are available, leading to distinct ignition probability for each release cause and size depending on pipeline category, toxicity/flammability of the material, pipeline size, pressure inside the pipeline, and release size.

The algorithm of models in PipeFAIT is shown in Fig. 37, and a screenshot of the front page showing the input parameters and output frequencies/ignition probabilities is shown in Fig. 38.

PipeFAIT produces tables/figures in *MS-Excel* format. A detailed report including all input parameters and results, as well as some suggestions for mitigation measures (depending on the dominant mode/cause of failure) is generated for the given pipeline.

TABLE 14 Example of Original Frequency Data Before Modification

	Original Frequency							
	Cause							
							Total	
Size (in.)	External Impact	Corrosion	Material/ Construction Defect	Incorrect Operations	Natural Forces	Others	Freq.	%
0.5 in.	2.7E−05	2.7E−05	2.9E−05	3.2E−06	1.2E−05	1.4E−05	1.1E−04	53%
2.0 in.	1.1E−05	1.1E−05	1.2E−05	1.3E−06	5.2E−06	5.7E−06	4.7E−05	22%
6.0 in.	5.6E−06	5.6E−06	6.0E−06	6.6E−07	2.6E−06	2.8E−06	2.3E−05	11%
FBR	6.9E−06	6.8E−06	7.4E−06	8.1E−07	3.2E−06	3.5E−06	2.9E−05	14%
Total	5.1E−05	5.0E−05	5.4E−05	6.0E−06	2.3E−05	2.6E−05	2.1E−04	100%
% of total	24%	24%	26%	3%	11%	12%	100%	

TABLE 15 Modified Frequency Using FFo Shown in Table 13 and Original Frequency in Table 14

Updated Frequency								
	Cause							
							Total	
Size (in.)	External Impact	Corrosion	Material/ Construction Defect	Incorrect Operations	Natural Forces	Others	Freq.	%
0.5 in.	1.2E−04	2.7E−07	5.2E−06	9.5E−06	1.3E−05	1.4E−05	1.7E−04	53%
2.0 in.	5.2E−05	1.1E−07	2.2E−06	4.0E−06	5.3E−06	5.7E−06	6.9E−05	22%
6.0 in.	2.6E−05	5.6E−08	1.1E−06	2.0E−06	2.6E−06	2.8E−06	3.5E−05	11%
FBR	3.2E−05	6.8E−08	1.3E−06	2.4E−06	3.2E−06	3.5E−06	4.2E−05	14%
Total	2.3E−04	5.0E−07	9.7E−06	1.8E−05	2.4E−05	2.6E−05	3.1E−04	100%
% of total	75%	0%	3%	6%	8%	8%	100%	

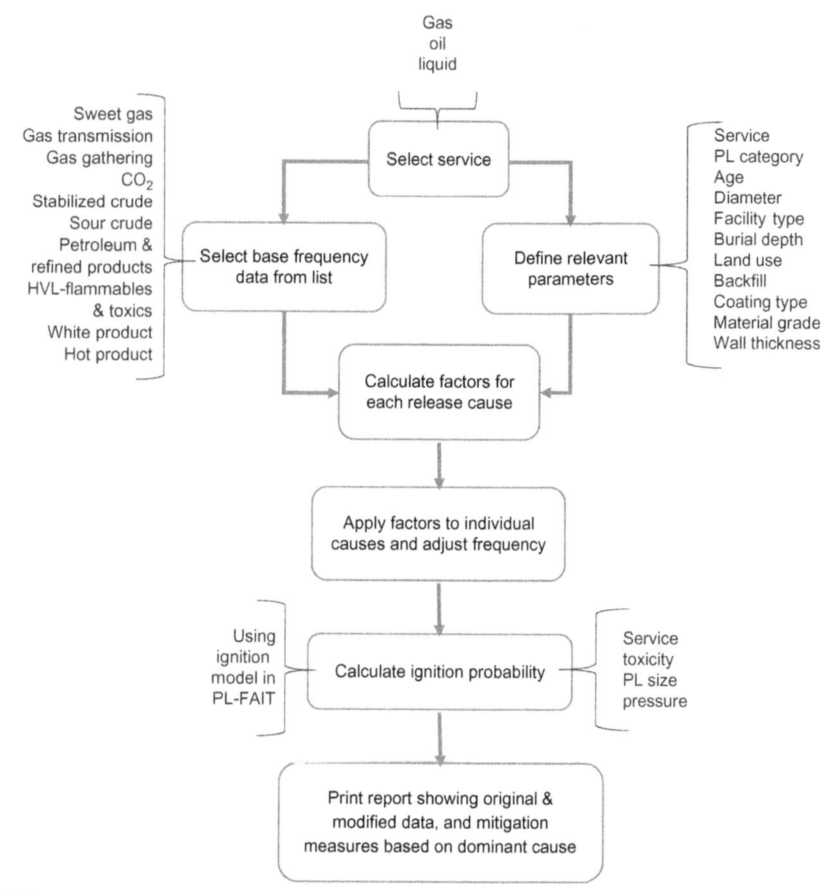

FIG. 37 PipeFAIT model algorithm.

Fig. 39 shows the ignition probability chart produced by PipeFAIT. The report also shows the relative pipeline risk index, defined as the pipeline's potential failure rate relative to average pipeline statistics in the industry within similar pipeline categories. An example of such report is shown in Fig. 40.

SUMMARY

A summary of all pipeline failure databases is presented, and a new pipeline failure assessment and integrity tool was developed to evaluate pipeline failure modes and leak frequencies for use in integrity and risk assessment and mitigation. The tool (PipeFAIT) is based on pipeline leak databases in the United States, Europe, and the United Kingdom. Overall, > 11,000 pipeline incidents were analyzed over the last four decades, and the results are used to develop the tool. Table 16 summarizes all the data used in the frequency analysis presented in this chapter.

LP Name	Pipeline XYZ

Gas Parameter	Value
Pipeline Category	Gas
Pipeline Service	Clean Gas
Pipeline Age	50 - 30 yrs
Pipeline Size	12 - 24 inch
Wall thickness	10 mm -15 mm
External Coating	Bitumen
Area Classification	Suburban
Facility Type	Underground Pipe
Back Fill	Sand
Burial Depth	< 80 cm Burial Depth
Material Grade	CRA (Corrosion Resistant Alloy)
Pipeline Pressure	<=35 bar
PL Contains Toxic?	No

Modified Pipeline Frequency									
	Cause								
Size (inch)	External Impact	Corrosion	Material/ Construction Defect	Incorrect Operations	Natural Forces	Others	Total		
							Freq.	%	TRUE
0.5 inch	3.2E-05	0.0E+00	7.3E-06	3.5E-06	7.0E-06	0.0E+00	4.9E-05	26%	TRUE
2.0 inch	1.3E-05	0.0E+00	3.0E-06	1.4E-06	2.9E-06	0.0E+00	2.1E-05	11%	TRUE
6.0 inch	3.4E-05	0.0E+00	7.7E-06	3.7E-06	7.4E-06	0.0E+00	5.2E-05	28%	TRUE
FBR	4.1E-05	0.0E+00	9.5E-06	4.5E-06	9.1E-06	0.0E+00	6.4E-05	34%	TRUE
Total	1.2E-04	0.0E+00	2.8E-05	1.3E-05	2.7E-05	0.0E+00	1.9E-04	100%	TRUE
% of Total	64%	0%	15%	7%	14%	0%	100%	TRUE	
	TRUE	TRUE	TRUE	TRUE	TRUE	TRUE	TRUE		

FIG. 38 See legend on opposite page.

PL Specific Ignition Prob. %	
Release Size	Ignition Prob. of Gas Release
0.5 inch	4%
2.0 inch	2%
6.0 inch	10%
FBR	12%

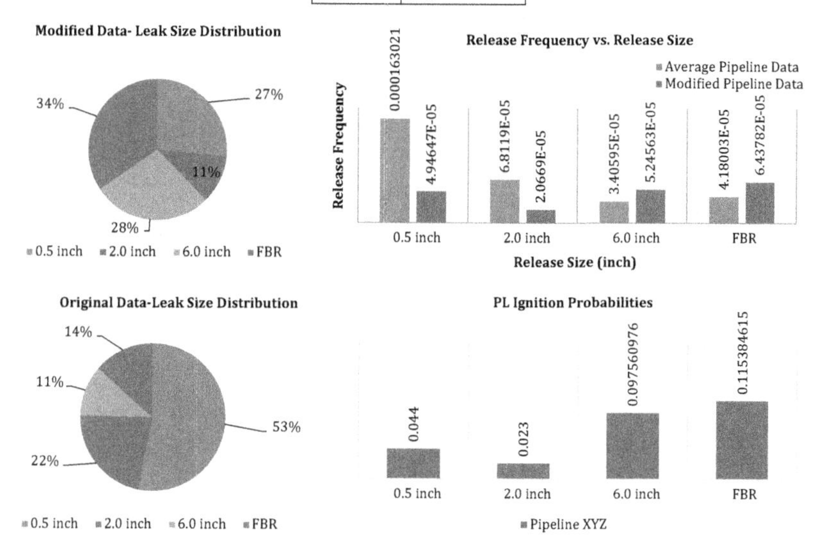

FIG. 38 PipeFAIT front page showing input parameters and output frequencies and ignition probabilities.

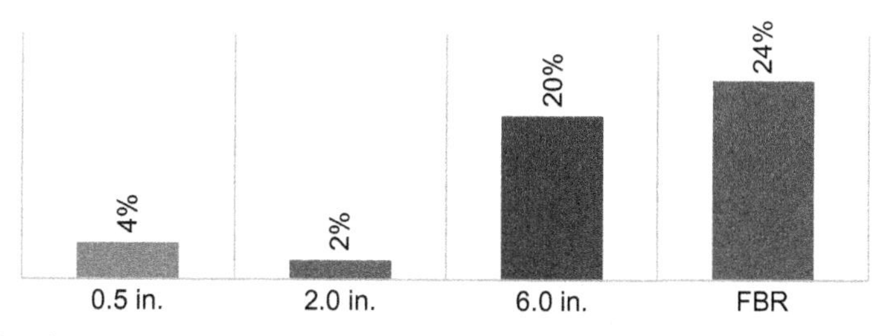

FIG. 39 PipeFAIT ignition chart per release size.

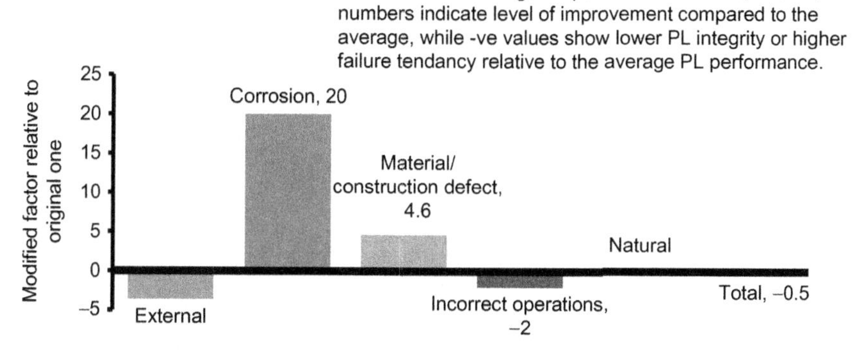

FIG. 40 PipeFAIT pipeline risk index chart.

TABLE 16 Summary of Leak Frequency for All Different Databases

Period	Category	Time Interval	No. of Incidents	Exposure (km/year)	Leaks (per km/year)
EGIG	Total period	1974–2013	1179	3.84E+06	3.07E−04
	Last 20 years	1994–2013	426	2.40E+06	1.78E−04
Canadian BC Oil and Gas Commission	Total data	2009–12	148	1.51E+05	9.8E−04
	Crude	2012	6	2.0E+03	3.0E−3
	Natural gas—sour	2012	7	1.3E+04	5.5E−04
	Natural gas—sweet	2012	4	1.8E+04	2.2E−04
CONCAWE	Total inventory	1971–2013	524	1.14E+06	4.60E−04
		1993–2013	230	7.3E+05	3.3E−04
	Crude	1971–2013	157	3.94E+05	4.0E−04
		1993–2013	52	2.2E+05	2.5E−04
	White product	1971–2013	297	7.19E+05	2.7E−03
		1993–2013	173	5.0E+05	3.6E−04
	Hot product	1971–2013	70	2.60e+04	4.1E−06
		1993–2013	5	7.6E+03	6.9E−04

UKOPA	Total period	1964–2013	187	8.49E+05	2.20E−04
	Last 20 years	1994–2013	33	4.30E+05	7.70E−05
	Last 5 years	2009–13	12	1.14E+05	1.00E−04
PHMSA	Total liquid	2004–13	4057	2.84E+06	1.43E−03
	Crude oil		1914	8.5E+05	2.25E−03
	Refined product		1481	1.0E+06	1.46E+03
	HVL flammable-toxic		556	9.0E+05	6.24E−04
	CO_2		47	6.8E+04	6.87E−04
	Gas transmission	1994–2013	2182	1.04E+07	2.10E−04
	Gas gathering		227	758,402	2.99E−04
	Gas distribution		2573	6.26E+07	4.11E−05

The assessment conducted in this chapter shows that failure mode and cause of a given pipeline depends on several factors including the design, operating, and environmental parameters. These factors include the pipeline material of construction (grade), wall thickness, operating pressure, service material, backfill medium/material, age, coating, pipeline size, and other relevant parameters. More information on specific failure mode is available in the literature [10,11].

PipeFAIT predicts the failure mode and patterns in terms of failure rate distribution by the size of leak and its causes. This allows for proper mitigation measures to be defined for most relevant failure cause. Ignition probabilities for pipeline failures were also analyzed and are predicted by this tool for each pipeline leak depending on the leak characteristics.

The analysis presented in this chapter allows for accurate and consistent evaluation of pipeline failure modes and rates that reflect the actual parameters of the pipeline; allows for evaluating specific failure rates for different segments of the pipeline where the design, operating, and environmental parameters change; and presented results in a format ready for use in pipeline risk assessments.

REFERENCES

[1] EGIG, 9th Report of the European Gas Pipeline Incident Data Group (Period 1970–2013), Doc. Number EGIG 14.R.0403, February, 2015.

[2] CONCAWE, Performance of European Cross Country Oil Pipeline, Statistical Summary of Reported Spillages in 2013 and Since 1971, Report No. 4/15, May, 2015.

[3] UKOPA, Pipeline Product Loss Incidents and Fault Report (1962–2013), UKOPA/14/0031, December, 2014.

[4] BC Oil and Gas Commission, Pipeline Performance and Activity Report, 2012.

[5] USA DOT PHMSA database, http://www.phmsa.dot.gov/pipeline/library/data-stats (accessed 2015–2016).

[6] Canadian National Energy Board, Focus on Safety and Environment: A Comparative Analysis of Pipeline Performance 2000–2008, August, 2010.

[7] CONCAWE, Performance of European Cross Country Oil Pipeline, Statistical Summary of Reported Spillages in 2014 and Since 1971, Report No. 7/16, June, 2016.

[8] K.E. Abhulimen, A.A. Susu, Liquid pipeline leak detection system: model development and numerical simulation, Chem. Eng. J. 97 (1) (2004) 47–67.

[9] P.-S. Murvay, I. Silea, A survey on gas leak detection and localization techniques, J. Loss Prev. Process Ind. 25 (6) (2012) 966–973.

[10] A. Cosham, K.A. Macdonald, Best practice for the assessment of defects in pipelines—corrosion, Eng. Fail. Anal. 14 (7) (2007) 1245–1265.

[11] C. Lam, Statistical Analyses of Historical Pipeline Incident Data with Application to the Risk Assessment of Onshore Natural Gas Transmission Pipelines, Master Thesis, Department of Civil and Environmental Engineering, University of Western Ontario.

FURTHER READING

[12] K.A. Macdonald, A. Cosham, Best practice for the assessment of defects in pipelines—gouges and dents, Eng. Fail. Anal. 12 (5) (2005) 720–745.

[13] S.A. Shipilov, I. Le May, Structural integrity of aging buried pipelines having cathodic protection, Eng. Fail. Anal. 13 (7) (2006) 1159–1176.

[14] A.P. Teixeira, C. Guedes Soares, T.A. Netto, S.F. Estefen, Reliability of pipelines with corrosion defects, Int. J. Press. Vessel. Pip. 85 (4) (2008) 228–237.

[15] K.A. Macdonald, A. Cosham, C.R. Alexander, P. Hopkins, Assessing mechanical damage in offshore pipelines—two case studies, Eng. Fail. Anal. 14 (8) (2007) 1667–1679.

[16] M.D. Pandey, Probabilistic models for condition assessment of oil and gas pipelines, NDT & E Int. 31 (5) (1998) 349–358.

[17] C.I. Ossai, B. Boswell, I.J. Davies, Pipeline failures in corrosive environments—a conceptual analysis of trends and effects, Eng. Fail. Anal. 53 (2015) 36–58.

[18] C.R.F. Azevedo, Failure analysis of a crude oil pipeline, Eng. Fail. Anal. 14 (6) (2007) 978–994.

[19] T. Berstad, C. Dørum, J.P. Jakobsen, S. Kragset, H. Li, H. Lund, A. Morin, S.T. Munkejord, M.J. Mølnvik, H.O. Nordhagen, E. Østbya, CO_2 pipeline integrity: a new evaluation methodology, Energy Procedia 4 (2011) 3000–3007.

[20] J.L. Alamilla, E. Sosa, C.A. Sánchez-Magaña, R. Andrade-Valencia, A. Contreras, Failure analysis and mechanical performance of an oil pipeline, Mater. Des. 50 (2013) 766–773.

[21] Y. Sun, L. Ma, J. Morris, A practical approach for reliability prediction of pipeline systems, Eur. J. Oper. Res. 198 (1) (2009) 210–214.

Chapter 6

Pipeline Quantitative Risk Assessment

Risk calculations for pipelines require both consequence and frequency to be calculated first, as described in Chapters 4 and 5. Once the consequences and frequencies are calculated, they are combined to assess the risk of specific scenario outcome, and then, once the risk of all scenario outcomes is assessed, it is calculated and added up to estimate the total risk from the pipeline. In this chapter, a detailed description of pipeline risk calculations is given with examples to demonstrate the risk. The literature has indicated that there is a need for simple model for pipeline risk assessment [1]. A detailed description of risk assessments and how they apply to pipeline is given in the literature [1–14].

SCENARIO OUTCOMES FOR PIPELINES

Typical pipeline hazards were summarized in Chapter 2, and Fig. 1 summarizes the potential scenario outcomes. This figure is shown below for demonstration purposes. Multiple release sizes can cause loss of containment from the pipeline, and all of these cases should be assessed. Since it is difficult to have a continuous spectrum of releases assessed in the risk assessment, a discreet release size distribution is assumed. Typically, small, medium, large, and full-bore rupture (FBR) cases are used to represent the entire spectrum of release sizes. Generally, these can be represented with ½ in., 2 in., 6 in., and full pipeline size, respectively, as shown in Chapter 5.

Release sizes are then used to calculate discharged rate of material from the pipeline as described in Chapter 4. Release frequency of different release sizes is estimated using the approach described in Chapter 5. Conditional modifiers are then estimated following the approach described in Chapter 5 or other applicable methodologies. After all the information is prepared, it is then summarized in an event tree similar to the example shown above.

The extent of the damage from different hazards is established for each scenario outcome of each release size, following the approach described in Chapter 4. Then, people's vulnerabilities to different hazard types and limits are estimated based on that. The pipeline route is then superimposed on the map (or aerial view) of the site showing all impacted populations and facilities. Release scenarios of hazardous cloud/impact boundaries are then added to the map along

Cross Country Pipeline Risk Assessments and Mitigation Strategies
https://doi.org/10.1016/B978-0-12-816007-7.00006-8

Event	Leak size	Early ignition	Delayed ignition	Congestion/ confinement	Outcome scenario	Probability of events	
Pipeline leak 1.0E-04	0.5 in. 0.5	Yes 0.01			Jet fire 5.00E-07	5.0E-05	1.0E-04
		No 0.99	Yes 0.1	No 0.9	Flash fire 4.50E-06		
				Yes 0.1	Explosion 5.00E-07		
			No 0.89		Toxic impact 4.45E-05		
	2.0 in. 0.25			Same as above		2.5E-05	
	6.0 in. 0.15			Same as above		1.5E-05	
	FBR 0.1			Same as above		1.0E-05	

FIG. 1 Illustration of event tree for pipelines.

the pipeline route and rotated in all different directions following the wind rose of the site's weather to show the 360° impact of all potential releases. The potential impact in terms of portion of population exposed to certain vulnerabilities is estimated. This is then combined with frequencies and conditional modifiers to calculate the risk, which is then summed for all cases to calculate the total risk posed by the pipeline for certain communities or groups of population.

FREQUENCY'S CONDITIONAL MODIFIERS

Once the release frequency of a pipeline is estimated, the risk associated with different hazardous events associated with the pipeline can be calculated. But the probability of each event has to be assessed and applied to adjust the original total frequency first as shown in Fig. 1. Conditional modifiers are the conditions that modify the original release to produce the final outcome. For example, a release from a pipeline has to form a flammable cloud that can ignite, the wind speed and direction should be right/suitable to blow the cloud into a specific direction of ignition, and the cloud would ignite. The release size should be big enough to form the large cloud needed to reach the ignition source and impact the people with high enough impact to cause damage (injury or fatality). People have to be present at the time the cloud ignites, in order to be affected. As such, the probability of a person in a given location that can be impacted by a fire from a pipeline release becomes as shown below:

$$P_{\text{total}} = P_{\text{rs}} \times P_{\text{ws}} \times P_{\text{wd}} \times P_{\text{ig}} \times P_{\text{pp}}$$

where P_{total} is the total probability; P_{rs} is the release size probability (per distribution in Chapter 5); P_{ws} is the probability of wind speed being enough to form large cloud that can reach people locations; P_{wd} is the probability that the wind direction is drifting the cloud toward people locations; P_{ig} is the probability that the release will ignite; and P_{pp} is the probability that people will be present in that specific location at the time of the release.

There could be other factors that should be considered besides the ones mentioned above as well. The total probability is then combined with the pipeline total release frequency and with the vulnerability of the specific hazard limit to decide whether people will be at risk from this specific event and to calculate its contribution to the overall risk, according to the formula below:

$$R = F_{pipeline} \times P_{total} \times V_{event}$$

where R is the risk from specific event type; $F_{pipeline}$ is the total pipeline release frequency (per distribution in Chapter 5); and V_{event} is the person's vulnerability due to specific hazardous type and limit.

Once risk of all different cases shown in the event tree is calculated, they are added together to estimate the total risk levels as risk is additive. Most conditional modifiers were discussed in the previous chapters, except for wind directions, which are presented in the next section.

WIND DIRECTION AND PROBABILITIES

Weather conditions change almost on a continuous bases. The wind direction and speed can vary significantly throughout the year. Because a release from a pipeline can occur anytime in the year and in any direction, the average wind speeds and directions throughout the year should be evaluated. As such, wind rose for the site weather should be prepared and used in the risk assessment of the pipeline. Typically, weather data from the site are collected for a long period of time (around 3–5 years) and used to prepare average wind speed directions and probabilities. Typical wind rose data are shown in Figs. 2 and 3 for a 16-direction pattern. Other patterns can be used such as 12 and 4 patterns, but they are not typically used in pipeline risk assessment as they do not have enough resolution to map the risk satisfactorily.

Wind rose data can be prepared using typical worksheet packages such as MS Excel, but there are also professional software packages available to assess data as well.

Toxic and/or flammable cloud formed from released material following pipeline leak has to be rotated with the wind rose directions at the same probabilities shown in the wind rose, and the impacted facilities/population groups will be accounted for in the risk assessment accordingly. Fig. 4 demonstrates the concept.

In the example above, a leak from pipeline *XYZ* forms a cloud that drifts toward population areas 1, 2, and 3. As seen, the cloud covers the portions of

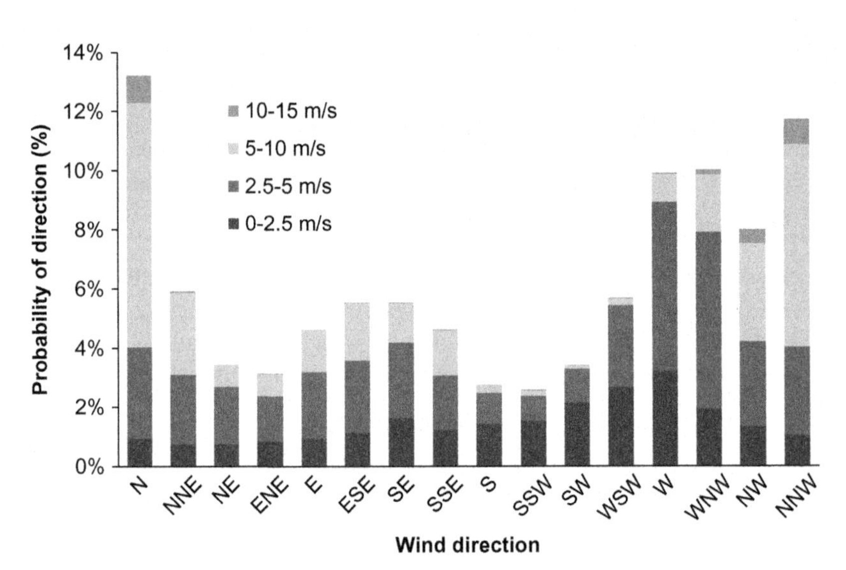

FIG. 2 Example of wind rose data used for risk assessments.

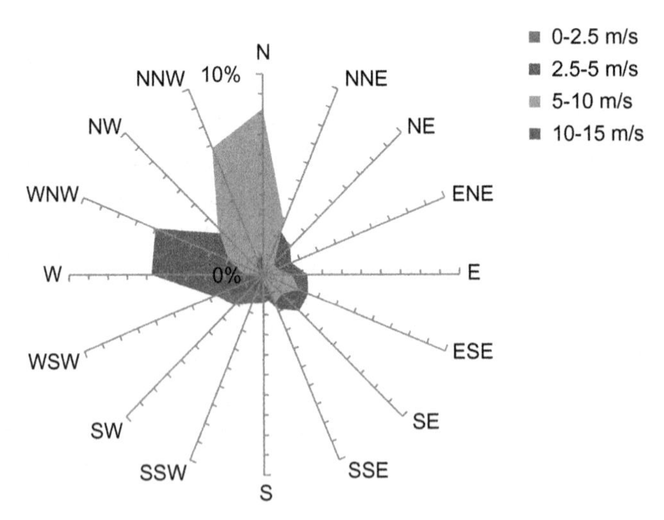

FIG. 3 Wind rose data shown on wind rose chart.

population area 1 (PA1) and population area 2 (PA2), but it does not impact population area 3 (PA3). Assuming that population is distributed evenly throughout each population area, then the portion of people impacted would be proportional to the portion of the areas covered by the different flammable limits.

For PA1, as shown in the example above, ¼ of the area is within UFL limit, ½ of the area is within LFL limits, and almost ¾ of PA1 area is within ½ LFL. For PA3, almost ⅔ of the area is impacted by ½ LFL cloud limit. Therefore, it can be estimated that ¼ of the population in PA1 is expected to have 100%

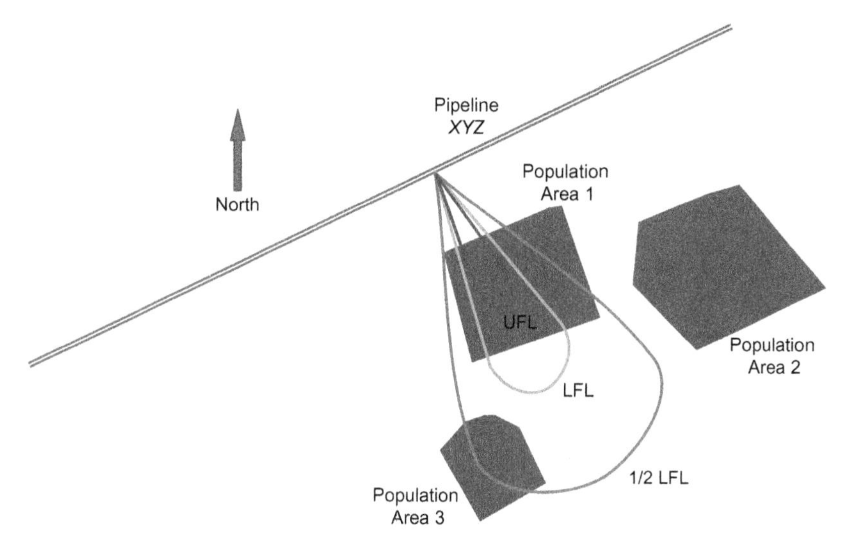

FIG. 4 Pipeline risk calculation example.

vulnerability (i.e., fatality probability) in case of flash fire (within UFL); roughly $\frac{1}{2}$–$\frac{1}{4}$ of PA1 will be at 70% fatality chance, and $\frac{3}{4}$–$\frac{1}{2}$ of PA1 population and $\frac{2}{3}$ of PA3 population are at risk of 10% fatality chance from flash fire as shown above. None of the population in PA2 is at risk in this case at all. The values used in this example are for demonstration purposes only.

However, these vulnerabilities will not occur every time there is a release, because the cloud can drift in other directions based on release direction and wind rose data. So, the vulnerabilities will be multiplied by the probability that the wind is going in that direction (south in this case). Based on the wind rose example shown above, this value will be 2.9%. The cloud will then be rotated in the other 15 directions of the wind rose, and the impact on all population areas and groups will be calculated in the same manner. Another factor to be considered is the probability of ignition that can cause flash fire (as shown in the event tree shown earlier in this chapter).

All other hazards such as thermal radiation from fires and toxic impact if any will be calculated in the same manner. Risk to each group is calculated from all scenario outcomes and then summed together to estimate the total risk from the pipeline to the community as demonstrated in Fig. 5. This is called the total societal risk. Other important risk measure is called individual risk and geographic risk. These will be discussed in the next section.

RISK MEASURES

Risk can be calculated and presented in different formats, sometimes called risk measures. In this book, two different measures will be presented as discussed in the following sections.

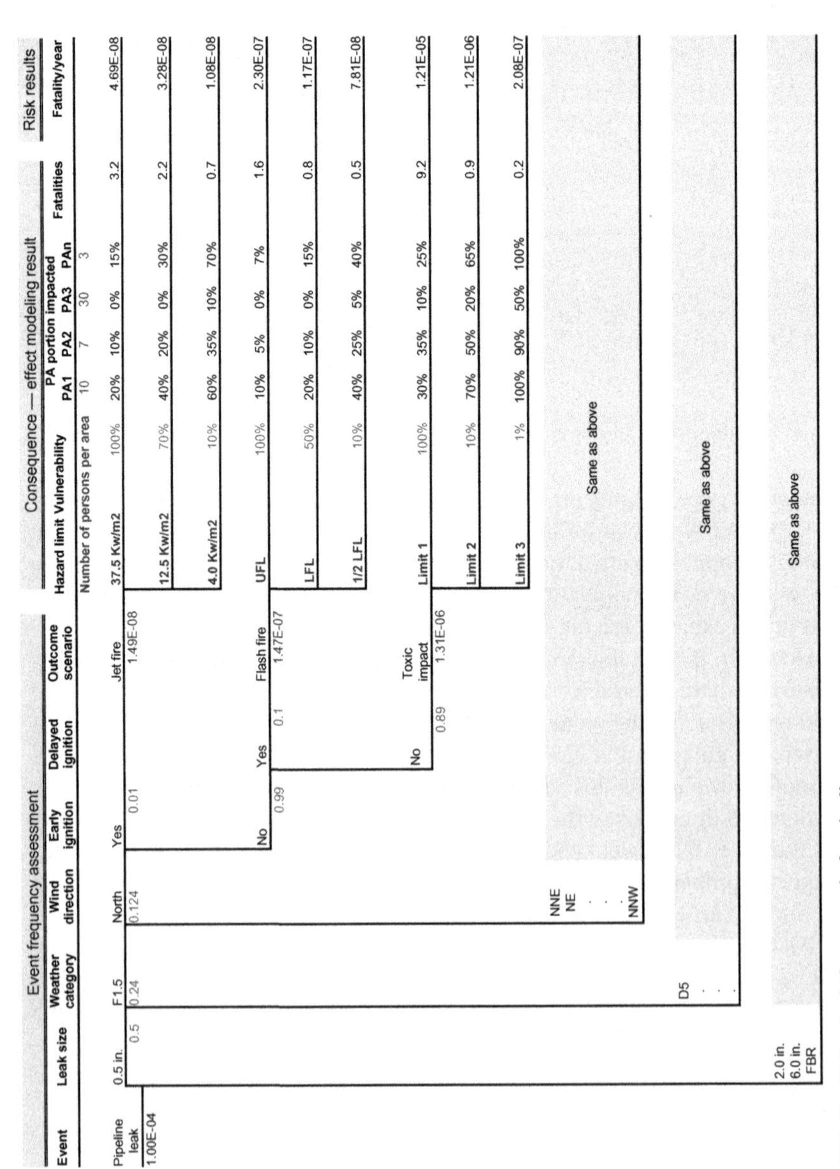

FIG. 5 Risk calculation example for pipelines.

Societal or Aggregate Risks

Societal risk is typically presented in the F-N curve, which plots the cumulative frequency of events that can cause N+ number of fatalities versus number of fatalities (such as data shown in Table 1). Typical example of F-N curve for pipelines is shown in Fig. 6. Societal risk for pipelines can be prepared for a specific community or per kilometer of the pipeline depending on the objective of the assessment. This is done for long pipelines that pass by several communities, which makes the total risk posed by the pipeline for all communities misleading. The choice of the appropriate measure is lift to the risk evaluator.

TABLE 1 Pipeline Societal Risk FN Data

Number of Fatalities	Frequency (event/year)	Cumulative Frequency (per year)
48	4.1E−05	4.1E−05
47	8.3E−05	1.2E−04
46	5.1E−06	1.3E−04
44	7.8E−05	2.1E−04
43	2.4E−05	2.3E−04
41	1.2E−05	2.4E−04
40	4.7E−05	2.9E−04
38	9.8E−06	3.0E−04
37	4.4E−05	3.4E−04
32	9.3E−05	4.4E−04
30	7.5E−05	5.1E−04
28	3.6E−05	5.5E−04
27	1.7E−05	5.6E−04
26	9.8E−05	6.6E−04
21	4.1E−05	7.0E−04
21	6.0E−07	7.0E−04
19	2.6E−05	7.3E−04
17	4.1E−07	7.3E−04
16	8.9E−05	8.2E−04
14	2.5E−06	8.2E−04
12	5.4E−05	8.8E−04

(Continued)

TABLE 1 Pipeline Societal Risk FN Data—cont'd

Number of Fatalities	Frequency (event/year)	Cumulative Frequency (per year)
11	1.1E−05	8.9E−04
9	5.1E−05	9.4E−04
8	6.4E−05	1.0E−03
6	7.7E−05	1.1E−03
5	1.9E−06	1.1E−03
4	4.2E−06	1.1E−03
3	9.9E−05	1.2E−03
2	6.1E−05	1.2E−03

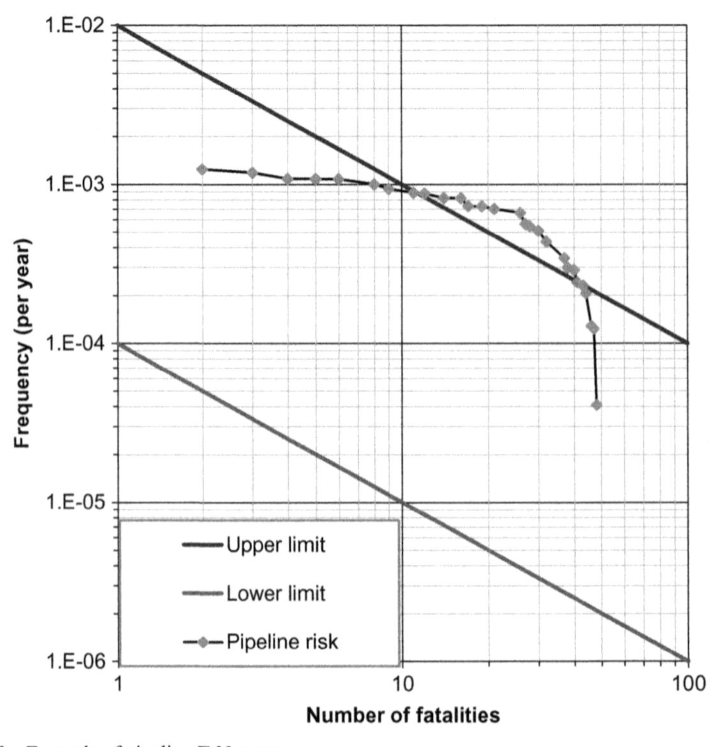

FIG. 6 Example of pipeline F-N curve.

Individual Geographic Risk

Individual geographic risk (IR) is calculated for the pipeline using consequence modeling and failure frequency along with all modifying factors such as ignition probabilities and wind directions, assuming that an individual is present all the time in the specific locations around the pipeline. The contribution of all types of hazards is accounted for, and the risk profile/decay is calculated starting from the pipeline and moving away. For the example given in Fig. 5, IR is calculated for thermal and flammable hazards as shown in Fig. 7. The figure is not necessarily a typical risk profile of natural gas pipeline, but is used only as an example. As shown, the risk starts high close to the pipeline and decay as the distance increases away from the pipeline. The graph also shows the risk contribution of the different types of hazards considered here (flammable and thermal). It shows that the risk is mainly dominated by the thermal hazard close to the pipeline and flammable hazard away from the pipeline.

Detailed calculations for this example are shown in the tables below. For simplicity, only one wind direction was assumed here, but in real cases, all wind directions shall be used per wind rose data. These calculations can be performed through a simple regular calculation sheet, but commercial software is available to conduct these calculations as well.

Table 2 summarizes the conditional modifiers and their values used in the risk calculations of the natural gas pipeline example presented here.

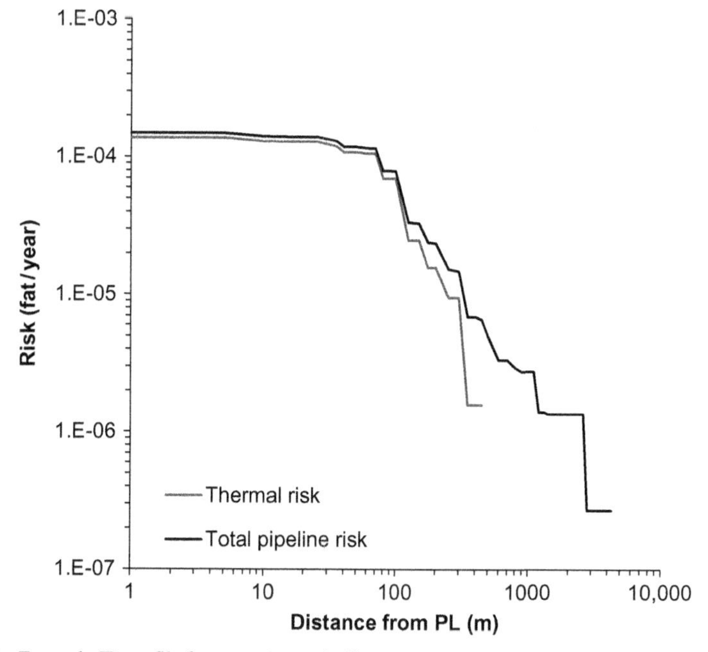

FIG. 7 Example IR profile for natural gas pipeline.

TABLE 2 Conditional Modifiers Used in Risk Calculations Example[a]

Variable	Parameter	Probability
Weather category	D5	60%
	F2	40%
Wind direction	South	15%
Flammable hazard	UFL	100%
	LFL	50%
	1/2 LFL	10%
Size distribution	0.5 in.	60%
	2.0 in.	25%
	6.0 in.	10%
	FBR	5%
PL frequency	Per year	1.00E−02
Late ignition	0.5	0.04%
	2	0.80%
	60	0.80%
	FBR	9.00%
Immediate ignition	0.5	1.0%
	2	6.2%
	60	6.2%
	FBR	21.0%
Thermal radiation	$\geq 37.5\,kW/m^2$	100%
	$12.5\,kW/m^2$	60%
	$4.0\,kW/m^2$	10%

[a] *Values used in this table are used for demonstration purposes and do not necessarily represent typical or default values. Vulnerabilities associated with different hazards limits can be found in the literature [22].*

Table 3 summarizes the consequence modeling results in terms of the distances reached by different hazard limits, which will be used to determine if a given population center or target is impacted by a given hazard level. Tables 4 and 5 show the impact of different hazard levels/limits with distance from the pipeline source for different release sizes and weather conditions. Table 6 shows the thermal hazard risk and combines that with the flammable risk values from Tables 4 and 5 to give the total risk distribution with distance as shown in Fig. 7.

TABLE 3 Natural Gas Pipeline Consequence Modeling Results

Release Size (in.)	Released Rate (kg/s)	Distance to Hazardous Limit (m)						Flame Length	Thermal Radiation Distance (m)		
		Weather Conditions, F2			Weather Conditions, D5						
		½ LFL	LFL	UFL	½ LFL	LFL	UFL	m	37.5	12.5	4
0.5	0.5	94	66	34	22	17	10	2	6	9	15
2	8	400	280	140	90	60	35	10	26	37	59
6	69	1300	890	430	280	190	100	29	76	111	173
24	405	4400	2660	1160	750	500	240	117	231	314	466

TABLE 4 Natural Gas Pipeline Risk From Flammable Hazard at D5 Weathers With Distances

Weather	Risk at a Given Distance at D5 Weather Condition											
Flammable Limit	UFL				LFL				1/2 LFL			
Release size	0.5	2	6	FBR	0.5	2	6	FBR	0.5	2	6	FBR
Distance	10	35	100	240	17	60	190	500	22	90	280	750
Size frequency distribution	60%	25%	10%	5%	60%	25%	10%	5%	60%	25%	10%	5%
Weather distribution	60%	60%	60%	60%	60%	60%	60%	60%	60%	60%	60%	60%
Wind direction	15%	15%	15%	15%	15%	15%	15%	15%	15%	15%	15%	15%
Flammable vulnerability	100%	100%	100%	100%	50%	50%	50%	50%	10%	10%	10%	10%
Late ignition	0%	1%	1%	9%	0%	1%	1%	9%	0%	1%	1%	9%
Pipeline release frequency	1.0E−02	1.0E−02	1.0E−02	1.0E−02	1.0E−02	1.0E−02	1.0E−02	1.0E−02	1.0E−02	1.0E−02	1.0E−02	1.0E−02
1	2.2E−07	1.8E−06	7.2E−07	4.1E−06								
2	2.2E−07	1.8E−06	7.2E−07	4.1E−06								
3	2.2E−07	1.8E−06	7.2E−07	4.1E−06								
4	2.2E−07	1.8E−06	7.2E−07	4.1E−06								
5	2.2E−07	1.8E−06	7.2E−07	4.1E−06								
10	2.2E−07	1.8E−06	7.2E−07	4.1E−06								

15		1.8E−06	7.2E−07	4.1E−06	1.1E−07							
20		1.8E−06	7.2E−07	4.1E−06					2.2E−08			
25		1.8E−06	7.2E−07	4.1E−06								
35		1.8E−06	7.2E−07	4.1E−06								
40			7.2E−07	4.1E−06		9.0E−07						
45			7.2E−07	4.1E−06		9.0E−07						
50			7.2E−07	4.1E−06		9.0E−07						
60			7.2E−07	4.1E−06		9.0E−07						
70			7.2E−07	4.1E−06						1.8E−07		
80			7.2E−07	4.1E−06						1.8E−07		
90			7.2E−07	4.1E−06						1.8E−07		
100			7.2E−07	4.1E−06								
125				4.1E−06			3.6E−07					
150				4.1E−06			3.6E−07					
175				4.1E−06			3.6E−07					
200				4.1E−06							7.2E−08	
250								2.0E−06			7.2E−08	
300								2.0E−06				
350								2.0E−06				

(Continued)

TABLE 4 Natural Gas Pipeline Risk From Flammable Hazard at D5 Weathers With Distances—cont'd

Weather	Risk at a Given Distance at D5 Weather Condition											
Flammable Limit	UFL				LFL				1/2 LFL			
400								2.0E−06				
450								2.0E−06				
500								2.0E−06				
600												4.1E−07
700												4.1E−07
800												
900												
4300												

TABLE 5 Natural Gas Pipeline Risk From Flammable Hazard at F2 Weathers With Distances

Weather	Risk at a Given Distance at F2 Weather Condition												Total Flash Fire IR (fatality/year)
Flammable Limit	UFL				LFL				1/2 LFL				
Release size	0.5	2	6	FBR	0.5	2	6	FBR	0.5	2	6	FBR	
Distance	34	140	430	1160	66	280	890	2660	94	400	1300	4400	
Size frequency distribution	60%	25%	10%	5%	60%	25%	10%	5%	60%	25%	10%	5%	
Weather distribution	40%	40%	40%	40%	40%	40%	40%	40%	40%	40%	40%	40%	
Wind direction	15%	15%	15%	15%	15%	15%	15%	15%	15%	15%	15%	15%	
Flammable vulnerability	100%	100%	100%	100%	50%	50%	50%	50%	10%	10%	10%	10%	
Late ignition	0%	1%	1%	9%	0%	1%	1%	9%	0%	1%	1%	9%	
Pipeline release frequency	1.0E−02	1.0E−02	1.0E−02	1.0E−02	1.0E−02	1.0E−02	1.0E−02	1.0E−02	1.0E−02	1.0E−02	1.0E−02	1.0E−02	
1	1.4E−07	1.2E−06	4.8E−07	2.7E−06									1.1E−05

(*Continued*)

TABLE 5 Natural Gas Pipeline Risk From Flammable Hazard at F2 Weathers With Distances—cont'd

Weather	Risk at a Given Distance at F2 Weather Condition												
Flammable Limit	UFL				LFL				1/2 LFL				Total Flash Fire IR (fatality/year)
2	1.4E−07	1.2E−06	4.8E−07	2.7E−06									1.1E−05
3	1.4E−07	1.2E−06	4.8E−07	2.7E−06									1.1E−05
4	1.4E−07	1.2E−06	4.8E−07	2.7E−06									1.1E−05
5	1.4E−07	1.2E−06	4.8E−07	2.7E−06									1.1E−05
10	1.4E−07	1.2E−06	4.8E−07	2.7E−06									1.1E−05
15	1.4E−07	1.2E−06	4.8E−07	2.7E−06									1.1E−05
20	1.4E−07	1.2E−06	4.8E−07	2.7E−06									1.1E−05
25	1.4E−07	1.2E−06	4.8E−07	2.7E−06									1.1E−05
35		1.2E−06	4.8E−07	2.7E−06	7.2E−08								1.1E−05
40		1.2E−06	4.8E−07	2.7E−06	7.2E−08								1.0E−05
45		1.2E−06	4.8E−07	2.7E−06	7.2E−08								1.0E−05
50		1.2E−06	4.8E−07	2.7E−06	7.2E−08								1.0E−05
60		1.2E−06	4.8E−07	2.7E−06	7.2E−08								1.0E−05
70		1.2E−06	4.8E−07	2.7E−06					1.4E−08				9.3E−06

80		1.2E−06	4.8E−07	2.7E−06					1.4E−08				9.3E−06
90		1.2E−06	4.8E−07	2.7E−06					1.4E−08				9.3E−06
100		1.2E−06	4.8E−07	2.7E−06									9.2E−06
125		1.2E−06	4.8E−07	2.7E−06									8.8E−06
150			4.8E−07	2.7E−06		6.0E−07							8.2E−06
175			4.8E−07	2.7E−06		6.0E−07							8.2E−06
200			4.8E−07	2.7E−06		6.0E−07							7.9E−06
250			4.8E−07	2.7E−06		6.0E−07							5.9E−06
300			4.8E−07	2.7E−06						1.2E−07			5.3E−06
350			4.8E−07	2.7E−06						1.2E−07			5.3E−06
400			4.8E−07	2.7E−06						1.2E−07			5.3E−06
450				2.7E−06			2.4E−07						5.0E−06
500				2.7E−06			2.4E−07						5.0E−06
600				2.7E−06			2.4E−07						3.3E−06
700				2.7E−06			2.4E−07						3.3E−06
800				2.7E−06			2.4E−07						2.9E−06
900				2.7E−06							4.8E−08		2.7E−06
1000				2.7E−06							4.8E−08		2.7E−06
1100				2.7E−06							4.8E−08		2.7E−06

(Continued)

TABLE 5 Natural Gas Pipeline Risk From Flammable Hazard at F2 Weathers With Distances—cont'd

Weather	Risk at a Given Distance at F2 Weather Condition												Total Flash Fire IR (fatality/year)
Flammable Limit	UFL				LFL				1/2 LFL				
1200								1.4E−06			4.8E−08		1.4E−06
1300								1.4E−06			4.8E−08		1.4E−06
1400								1.4E−06					1.4E−06
1500								1.4E−06					1.4E−06
1600								1.4E−06					1.4E−06
1700								1.4E−06					1.4E−06
1800								1.4E−06					1.4E−06
1900								1.4E−06					1.4E−06
2000								1.4E−06					1.4E−06
2200								1.4E−06					1.4E−06
2400								1.4E−06					1.4E−06
2600								1.4E−06					1.4E−06
2800												2.7E−07	2.7E−07
4000												2.7E−07	2.7E−07
4300												2.7E−07	2.7E−07

TABLE 6 Natural Gas Pipeline Risk From Thermal Hazard and Total Risk With Distances

Weather	Thermal Impact Risk Levels and Total Risk Levels at a Given Distance													
Flammable Limit	**37.5**				**12.5**				**4**				**Total Thermal IR (fatality/year)**	**Total Individual Risk (fatality/year)**
Release size	0.5	2	6	FBR	0.5	2	6	FBR	0.5	2	6	FBR		
Distance	6	26	76	231	9	37	111	314	15	59	173	466		
Size frequency distribution	60%	25%	10%	5%	60%	25%	10%	5%	60%	25%	10%	5%		
Weather distribution	100%	100%	100%	100%	100%	100%	100%	100%	100%	100%	100%	100%		
Wind direction	15%	15%	15%	15%	15%	15%	15%	15%	15%	15%	15%	15%		
Flammable vulnerability	100%	100%	100%	100%	60%	60%	60%	60%	10%	10%	10%	10%		
Late ignition	1%	6%	60%	21%	1%	6%	60%	21%	1%	6%	60%	21%		
Pipeline release frequency	1.0E−02	1.0E−02	1.0E−02	1.0E−02	1.0E−02	1.0E−02	1.0E−02	1.0E−02	1.0E−02	1.0E−02	1.0E−02	1.0E−02		
1	8.6E−06	2.3E−05	9.0E−05	1.6E−05									1.4E−04	1.5E−04
2	8.6E−06	2.3E−05	9.0E−05	1.6E−05									1.4E−04	1.5E−04

(Continued)

TABLE 6 Natural Gas Pipeline Risk From Thermal Hazard and Total Risk With Distances—cont'd

Weather	Thermal Impact Risk Levels and Total Risk Levels at a Given Distance													
Flammable Limit	37.5				12.5				4				Total Thermal IR (fatality/year)	Total Individual Risk (fatality/year)
3	8.6E−06	2.3E−05	9.0E−05	1.6E−05									1.4E−04	1.5E−04
4	8.6E−06	2.3E−05	9.0E−05	1.6E−05									1.4E−04	1.5E−04
5	8.6E−06	2.3E−05	9.0E−05	1.6E−05									1.4E−04	1.5E−04
10		2.3E−05	9.0E−05	1.6E−05					8.6E−07				1.3E−04	1.4E−04
15		2.3E−05	9.0E−05	1.6E−05									1.3E−04	1.4E−04
20		2.3E−05	9.0E−05	1.6E−05									1.3E−04	1.4E−04
25		2.3E−05	9.0E−05	1.6E−05									1.3E−04	1.4E−04
35			9.0E−05	1.6E−05		1.4E−05							1.2E−04	1.3E−04
40			9.0E−05	1.6E−05						2.3E−06			1.1E−04	1.2E−04
45			9.0E−05	1.6E−05						2.3E−06			1.1E−04	1.2E−04
50			9.0E−05	1.6E−05						2.3E−06			1.1E−04	1.2E−04
60			9.0E−05	1.6E−05									1.1E−04	1.2E−04
70			9.0E−05	1.6E−05									1.1E−04	1.2E−04
80				1.6E−05			5.4E−05						7.0E−05	7.9E−05
90				1.6E−05			5.4E−05						7.0E−05	7.9E−05

100				1.6E−05			5.4E−05						7.0E−05	7.9E−05
125				1.6E−05							9.0E−06		2.5E−05	3.4E−05
150				1.6E−05							9.0E−06		2.5E−05	3.3E−05
175				1.6E−05									1.6E−05	2.4E−05
200				1.6E−05									1.6E−05	2.4E−05
250								9.5E−06					9.5E−06	1.5E−05
300								9.5E−06					9.5E−06	1.5E−05
350												1.6E−06	1.6E−06	6.9E−06
400												1.6E−06	1.6E−06	6.9E−06
450												1.6E−06	1.6E−06	6.5E−06
500													0.0E+00	5.0E−06
600													0.0E+00	3.3E−06
700													0.0E+00	3.3E−06
800													0.0E+00	2.9E−06
900													0.0E+00	2.7E−06
1000													0.0E+00	2.7E−06
1100													0.0E+00	2.7E−06
1200													0.0E+00	1.4E−06
1300													0.0E+00	1.4E−06
1400													0.0E+00	1.4E−06
1500													0.0E+00	1.4E−06

(Continued)

TABLE 6 Natural Gas Pipeline Risk From Thermal Hazard and Total Risk With Distances—cont'd

Weather	Thermal Impact Risk Levels and Total Risk Levels at a Given Distance													
Flammable Limit	37.5				12.5				4				Total Thermal IR (fatality/year)	Total Individual Risk (fatality/year)
1600													0.0E+00	1.4E−06
1700													0.0E+00	1.4E−06
1800													0.0E+00	1.4E−06
1900													0.0E+00	1.4E−06
2000													0.0E+00	1.4E−06
2200													0.0E+00	1.4E−06
2400													0.0E+00	1.4E−06
2600													0.0E+00	1.4E−06
2800													0.0E+00	2.7E−07
3000													0.0E+00	2.7E−07
3250													0.0E+00	2.7E−07
3500													0.0E+00	2.7E−07
3750													0.0E+00	2.7E−07
4000													0.0E+00	2.7E−07
4300													0.0E+00	2.7E−07

RISK TOLERANCE CRITERIA

Fig. 6 above also shows upper and lower risk limits. The upper limit indicates the intolerable risk limit to which no population should be exposed. The lower limit represents the broadly acceptable risk limits, and in between is the region of tolerable risk otherwise known as the region of ALARP. The limits shown in Fig. 6 are based on the criteria used by the UK-HSE guideline on risk to personnel working at the site. However, generally speaking, it is well accepted that off-site general public that is not directly involved in the operation of the hazardous facility (in this case the pipeline) should be exposed to lower risk because of several reasons highlighted below:

- The general public may not understand the hazard and risk associated with the pipeline running next to their communities.
- The public may not be prepared to react to emergencies should they occur, in the same manner as the operators of the pipeline who are trained and equipped for that.
- The general public might not be benefiting directly from the pipeline, and as such, they should not be expected to accept risk levels as high as the risk taken by operators who benefit from the pipeline operations directly. This is consistent with the general concept of positive/proportional relation between risk and reward.
- The general public would be invulnerably exposed to the risk against their choice versus the pipeline's operational personnel who choose to accept the risk vulnerably in exchange for the rewards/benefits. As such, general public risk limits would be lower.

Generally speaking, a general public risk limits would be one order of magnitude lower than the risk limits of the workers and operator's personnel. Therefore, the criteria shown in Fig. 6 would be lowered by one order of magnitude for both the upper and lower limits if the risk shown in the graph is for the general public not the pipeline workers. The upper limit would be 1.0E − 3 instead of 1.0E − 2 fatality/year, and the lower limit becomes 1.0E − 5 instead of 1.0E − 4 fatality/year. This is also similar to the risk criteria used in the "hazardous" industry in general.

It should be mentioned that if the regulators have specific risk tolerance criteria for the public and for the operator's personnel, then that should be used and respected. A summary of different risk tolerance criteria used in different industries/jurisdictions is given in the literature [15–23,26]. Pipeline operators might choose to use more stringent criteria compared with regulations applicable at the operations site, but not a less stringent one. For example, Netherland, UK, and Hong Kong authorities have defined criteria to be used. Some of these criteria are more stringent than others [19–21,23]. But not all other regulators/governments have defined criteria for societal risk. In that case, one can use the general death rate in the country as a guideline to set something acceptable. Some references suggest a value that is 1% of total mortality rate as an acceptable risk limit for involuntary risk [21]. This might be too conservative, and 10% could be used instead.

Fig. 8 shows the mortality rate for several countries in 2014 [25]. Most countries have a mortality rate above 1.0E − 3 fatality/year. As such, it might be acceptable to use a general public upper limit below 1.0E − 3 fatality/year, to make sure the pipeline operation does not expose the public to any risk higher than the normal everyday risk they are exposed to in their daily life. This concept represents the underlying philosophy for risk tolerance. One can suggest 1E − 5 fatalities per year as an acceptable limit of societal risk, for example, with maximum tolerable limits of 1E − 3 fatalities per year. Individual risk criteria should be one order of magnitude lower.

Some jurisdictions suggest using a risk criteria per kilometer of pipeline segments. Criteria in line with this are proposed in the literature [22]. The value is based on the UK and the Dutch criteria but starts at 1E − 4 fatalities per year with a log-log scale of − 1 and 1E − 2 fatalities per year with a log-log slope of − 2, respectively.

It is important to note that several risk assessments can be defined for different purposes and different types of assessments. For example, layer of protection analysis (LOPA) typically has a risk criteria that would be different from the societal criteria that is described here, because the intent of the LOPA criteria is to define feasibility of measures to protect the process and enhance the integrity of the risk mitigation measures such as emergency shutdown systems [24]. As such,

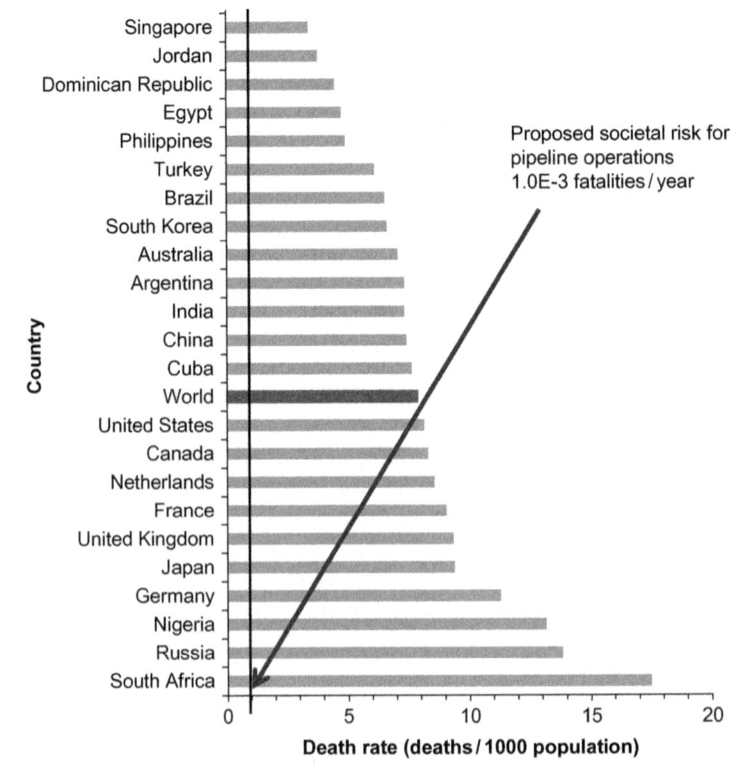

FIG. 8 Death rates for different societies.

the criteria presented in this book should only be used in the context described in this book and only for guidance rather than compliance with regulations.

LAND USE PLANNING

One of the most practical uses of IR calculations is to plan the use of lands around pipelines and reduce risk from pipelines to public facilities. In some countries, authorities and regulators have specific land-use planning (LUP) criteria that dictates the acceptable risk limits from pipeline to different uses of land. This links the risk level to the intended utilization of the land. The UK Health and Safety Executive (HSE) and others, for example, have specific criteria that links the risk level with the type of land use and density/nature of public centers around the pipelines and other facilities [27–30].

LUP criteria typically allow land use around pipeline based on risk level and type of land use/nature of public in these lands. Typically, the land use is categorized into the following:

- Industrial/agricultural.
- Low residential density.
- High residential density areas.
- Sometimes the criteria allow for exclusion zone where no public is allowed and vulnerable population zones (such as schools and hospitals).

The risk limits acceptable by authorities for different land use/population category are combined with the risk profile similar to the one shown in Fig. 7 to determine the areas used for different land-use purposes. Fig. 9 illustrates the concept.

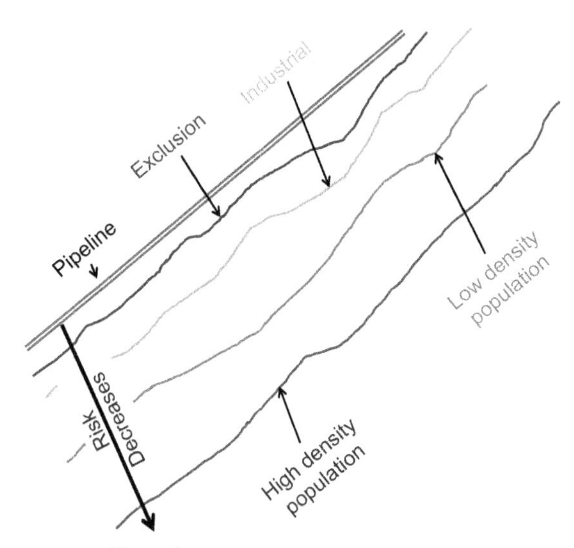

FIG. 9 Land-use concept illustration.

The exact value of the risk limit associated with each one of these land-use categories is dictated by authorities and regulations where the pipeline is operated. However, it is a good practice to limit the risk posed by the pipeline to the public to levels that do not significantly increase the general public risk as illustrated earlier in Fig. 8. If the regulations call for more stringent values, those shall be followed. Once the risk is calculated, then LUP zones can be easily established and checked against the applicable criteria for compliance and risk management purposes. General discussion on LUP and the proposed criteria that can be used is given in the literature [27–34].

RISK MITIGATION

If the risk calculations show that the risk is not acceptable or high, mitigation measures should be implemented to reduce the level of risk posed by the pipeline to the recipients. Since the risk is dependent on leak frequencies and consequences, it can be mitigated by either reducing the leak frequencies or mitigating the consequences of the release. A detailed list of pipeline risk reduction measures is given in Table 7. This list is not necessarily comprehensive, and not all measures are needed in every pipeline case. It is dependent on the case-by-case situation, but the list can be used for guidance purposes. Causes shown in this table are based on reported causes in PHMSA database discussed in Chapter 5.

TABLE 7 Pipeline Risk Mitigation Measures

Damage Category	Damage Cause	Potential Mitigation Measures
Corrosion	External	Cathodic protection
		External coating
		Backfill type
		Wall thickness
		Utc
		Increase burial depth
	Internal	Corrosion inhibitor
		Wall thickness
		Material grade
		Scraping
		ILI
		Corrosion monitoring programs
		Optimum velocity/operating conditions

TABLE 7 Pipeline Risk Mitigation Measures—cont'd

Damage Category	Damage Cause	Potential Mitigation Measures
Excavation damage	Operator/contractor excavation damage	Work permit
		PL markers and signs
		Work supervision by company personnel
		PL cover (e.g., concrete slab)
		Wall thickness
		Excavation procedures
	Third-party excavation damage	Fence
		PL corridors
		Physical barriers
		Increased burial depth
		PL cover (e.g., concrete slab)
		Wall thickness
		Regular patrol and survey
Incorrect operation	Damage by operator or operator's contractor	Proper operating procedures
		Work permits
		Supervision by company personnel
	Incorrect equipment	JSA, operating procedure and work permit
	Incorrect installation	Installation procedures
	Incorrect operation	Operating procedures
		Supervision/oversight of work
	Incorrect valve position	Lock open/lock close of critical valves
		Work permits
		Supervision of work
		Valve position alarms at control room
	Pipeline/equipment overpressured	ESD systems
		Fully rated pipeline
		Hipps
		Basic process control system (BPCS)
		Mechanical relief systems
		Operating procedures

(Continued)

TABLE 7 Pipeline Risk Mitigation Measures—cont'd

Damage Category	Damage Cause	Potential Mitigation Measures
Material/ weld/ equip. failure	Body of pipe	Proper welding procedures
		Work permits
		Scraping
		Inspection
		Preventative maintenance
	Compressor or compressor-related equipment	Preventative maintenance
		Inspection
		Operating procedures
	Construction, installation, or fabrication-related	Proper construction procedures
		Hydrotesting
		Mechanical completion certificate/process
	Defective or loose tubing/fitting	Preventative maintenance
		Inspection
	Environmental cracking-related	
	Failure of equipment body	Preventative maintenance
		Inspection
	Joint/fitting/component	Preventative maintenance
		inspection
	Malfunction of control/ relief equipment	Preventative maintenance
		Inspection
		Testing procedures
	Manufacturing-related	Hydrotesting
	Mechanical fitting	Preventative maintenance
		Inspection
	Nonthreaded connection failure	Inspection and preventative maintenance
	Other equipment failure	Inspection and preventative maintenance
	Other pipe/weld/joint failure	Preventative maintenance
		Inspection
		Scraping
		ILI

TABLE 7 Pipeline Risk Mitigation Measures—cont'd

Damage Category	Damage Cause	Potential Mitigation Measures
	Pump or pump-related equipment	Inspection, preventative maintenance, operating procedures, equipment integrity checks
	Ruptured or leaking seal/pump packing	
	Threaded connection/coupling failure	
	Threads stripped, broken pipe coupling	
	Valve	Preventative maintenance
		Inspection
Natural force damage	Earth movement	Proper support of PL
	Heavy rains/floods	Drainage around pipeline
	High winds	Pipeline support design
	Lightning	Design practices
	Other natural force damage	Proper seismic assessment
	Temperature	Design practices and pipeline burial
Other outside force damage	Electric arcing from other equipments/facilities	Proper electric equipment isolation
	Fire/explosion as primary cause	Proper spacing of PL/equipment
		Pipeline burial
	Fishing or maritime activity	Anchoring procedures
		Patrol of subsea pipeline routes
	Intentional damage	Security patrol
		Physical barriers
		Fences
	Maritime equipment or vessel adrift	Pipeline design/support
		Vessel operating procedures
	Vehicle not engaged in excavation	Vehicle access control
		Lifting operation plan
		Lift operator certification

In order to quantify the actual impact of these measures on pipeline risk, a risk reduction factor (RRF) must be used and evaluated based on actual data or common engineering sense/judgment. If data are available, it can be used in a manner similar to what was shown in Chapter 5 (frequency analysis) to evaluate the actual reduction factor. Otherwise, a qualitative assessment should be done. This section describes a methodology that can be used for that purpose.

Different risk mitigation measures are typically categorized as shown in Fig. 10, and their general effectiveness is proposed in the figure as well. For example, eliminating the hazard or adopting engineering measures to reduce the risk (such as ESD systems and isolation valves) is more effective than that of administrative measures (such as the following operating procedures that can be misinterpreted or ignored).

A multidisciplinary team of subject matter experts (SMEs) should conduct a workshop to define the different levels of effectiveness of these categories based on their understanding and level of trust in these measures at the field. Table 8 shows an example where the measures are rated on a scale from 0 to 10. The team should also define the status of the measures implemented in the specific pipeline case being analyzed and compare that to applicable general pipeline best practices and regulatory requirements as shown in Table 9. For example, if

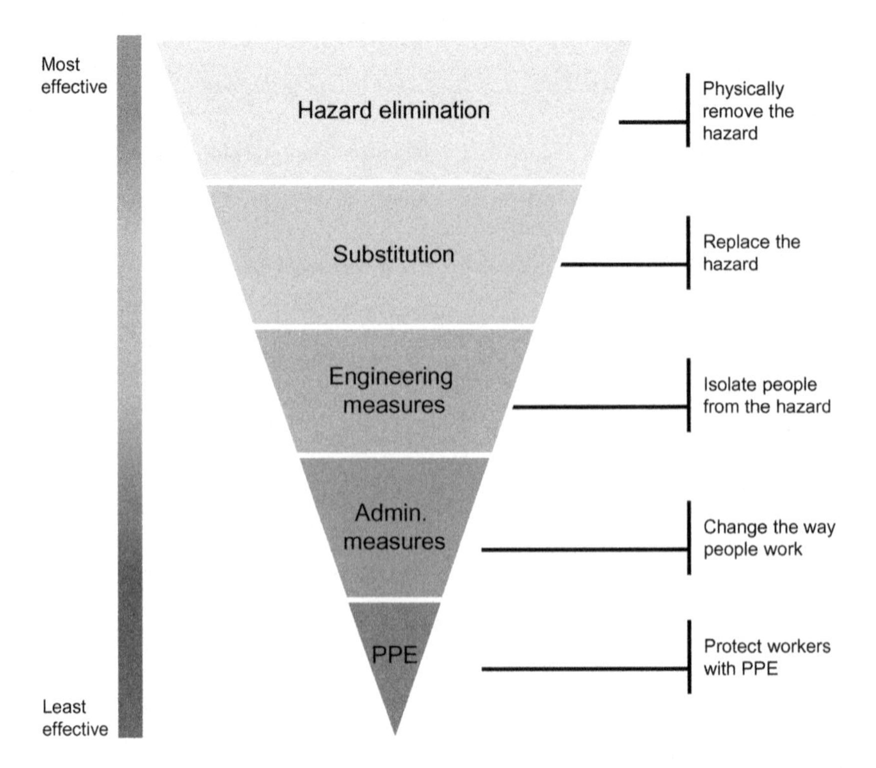

FIG. 10 Pipeline risk mitigation measures.

TABLE 8 Mitigation Measure Effectiveness

Category	Effectiveness
Hazard elimination	10.0
Substitution	7.5
Engineering control	5.0
Administration control	2.5
PPE	1.0

TABLE 9 Mitigation Measures Rank Versus Industry Best Practices

Company Practices Relative to Industry Best Practices	
Extremely better	5
Better	3
Similar	0
Worst	−3
Extremely worst	−5

operating procedures implemented in this specific pipeline operation are much better than the best practices, then their effectiveness would be high; otherwise, it would be low. It is proposed to rank the effectiveness from −5 (for extremely worse than industry best practices) to +5 (for extremely better than the best practices in the industry). But the team can use their own appropriate scale as long as that is defined and documented for future references and for consistency.

Once the ranking criteria shown in Tables 8 and 9 is fixed and accepted, then the different applicable measures shown in Table 7 can be evaluated in order to estimate their effectiveness in mitigating the risk. The table below illustrates an example of this concept being applied for corrosion mitigations. The overall factor shown in Table 10 is then multiplied by the frequency of leak due to corrosion, and the adjusted frequency is now used as a representative of pipeline leak frequency due to corrosion. Same concept is applied for all other causes of pipeline leaks such as external impact and operational errors.

If more appropriate data or justification for credit taken for risk mitigation is available, it should be used. The concept illustrated here in this section is to be used as a means of systematically measuring mitigation measure effectiveness in case of the lack of other more accurate/appropriate approaches. However, the proper application of this concept would still lead to adequate results.

TABLE 10 Corrosion Risk Mitigation Measures Evaluation

Mitigation Category	Mitigation Measure List	Company Practice	Justification/Notes	Factor (From −5 to +5)	Reduction/Increase in Frequency	Frequency Multiplier
Administrative control	Call in system	Better	Robust systems that is proved in field	3	2.5	$=1/(3\times 2.5)$
	Encroachment management process	Similar	Following regular general practices	0	2.5	No credit is taken (i.e., = 1)
	Excavation procedures	Better	Marked pipelines with concrete slab protections and clear signs	3	2.5	$=1/(3\times 2.5)$
	ILI	Better	In-line inspection conducted at higher frequency than required by best practices	3	2.5	$=1/(3\times 2.5)$
	Inspection	Worst	Unsupervised process that can be overdue	−3	2.5	$=3\times 2.5$
	JSA and work permit	Extremely worse	Not effective process that is bypassed easily	−5	2.5	$=5\times 2.5$
	Lifting operation plan	Similar	Following regular procedures	0	2.5	No credit is taken (i.e., = 1)
	Overall factor					Multiplication of all above factors (1/4.5)

PIPELINE RISK ASSESSMENT CASE STUDY

This case study is based on the risk assessment presented in Ref. [13]. The São Carlos-Porto Ferreira high-pressure natural gas (NG) pipeline in Brazil is buried and has a diameter of 200 mm and operates at a pressure of 35 bar_g. A risk assessment was performed, and the results were presented in Ref. [13]. The current methodology presented in this book was used to assess the risk and compare it with the results shown in the reference. A summary of the risk assessment performed per methodology presented in this book is summarized in this section.

- Input parameters and assumptions:
  - Pipeline diameter: 200 mm.
  - PL operating pressure 35 bar_g.
  - PL service: natural gas (methane assumed for simplifications).
  - Pipeline buried underground.
  - Ambient temperature is assumed.
- Consequence modeling conducted per methodology in Chapter 4. Dispersion analysis calculated assumes the following:
  - F2 and D5 weather conditions.
  - Low surface roughness.
  - Since this is NG, no VCE or UVCE is expected.
  - Crater size was calculated, and fatality was assumed imminent within the crater (100% vulnerability).
  - Only flammable and thermal hazards evaluated since there is no toxic material in the gas pipeline.
- Frequency data were used based on EGIG database for the overall frequency and UKOPA for the size distribution data:
  - Wind rose information is not available; a maximum wind direction probability of 15% in one direction is assumed.
  - Conditional modifiers were used per examples shown in this book.

The input parameters of the assessment are summarized in Table 11. The consequence modeling results are given in Tables 12 and 13. Crater size was found to be 9 m. The risk assessment results are given in terms of the individual risk (IR) profile versus the distance from the pipeline location in Fig. 11 to be compared with the results in the original reference.

TABLE 11 Conditional Modifiers and Vulnerabilities Used in São Carlos-Porto Ferreira PL Case Study

Variable	Parameter	Probability
Weather category	D5	50%
	F2	50%
Wind direction	South	15%

(Continued)

TABLE 11 Conditional Modifiers and Vulnerabilities Used in São Carlos-Porto Ferreira PL Case Study—cont'd

Variable	Parameter	Probability
Flammable hazard	UFL	100%
	LFL	50%
	½ LFL	10%
Size distribution	½ in.	79%
	2.0 in.	12%
	6.0 in.	5%
	FBR	4%
PL frequency	Per year	1.78E−04
Late ignition	½ in.	0.04%
	2.0 in.	0.80%
	6.0 in.	0.80%
	FBR	9.00%
Immediate ignition	½ in.	1.0%
	2.0 in.	6.2%
	6.0 in.	6.2%
	FBR	21.0%
Thermal radiation	≥37.5 kW/m^2	100%
	12.5 kW/m^2	60%
	4.0 kW/m^2	10%

TABLE 12 Flammable Dispersion Results for São Carlos-Porto Ferreira PL Case Study

Release Size (in.)	Released Rate (kg/s)	Distance to Hazardous Limit (m)					
		Weather Conditions: F2			Weather Conditions: D5		
		½ LFL	LFL	UFL	½ LFL	LFL	UFL
0.5	0.8	120	75	50	20	10	1
2	13	480	340	200	120	75	50
6	150	1680	1060	580	340	250	150
8	204	2400	1540	770	480	340	200

TABLE 13 Thermal Consequence Modeling Results for São Carlos-Porto Ferreira PL Case Study

Release Size (in.)	Flame Length (m)	Thermal Radiation Distance (m)		
		37.5 kW/m^2	12.5 kW/m^2	4 kW/m^2
0.5	2	8	11	18
2	10	30	45	72
6	29	99	149	241
8	39	120	179	286

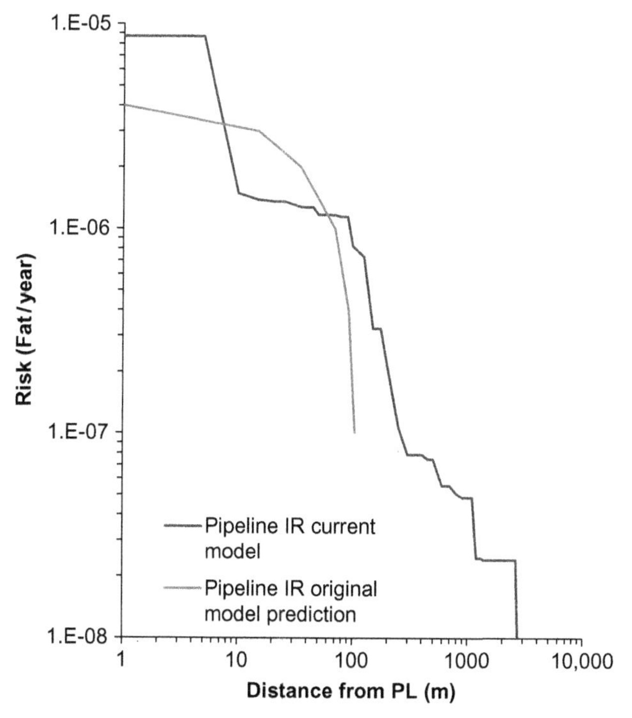

FIG. 11 IR profile for São Carlos-Porto Ferreira PL case study.

As shown in Fig. 11, the risk assessment results are very similar for both the original model and the simplified model presented in this book (referred to as the current model). The current model is simple and easy to use, where risk assessment can be performed quickly without consuming a lot of time. Yet, the results are similar to the more complicated/time-consuming models that do not necessary produce different results. The current model is slightly more

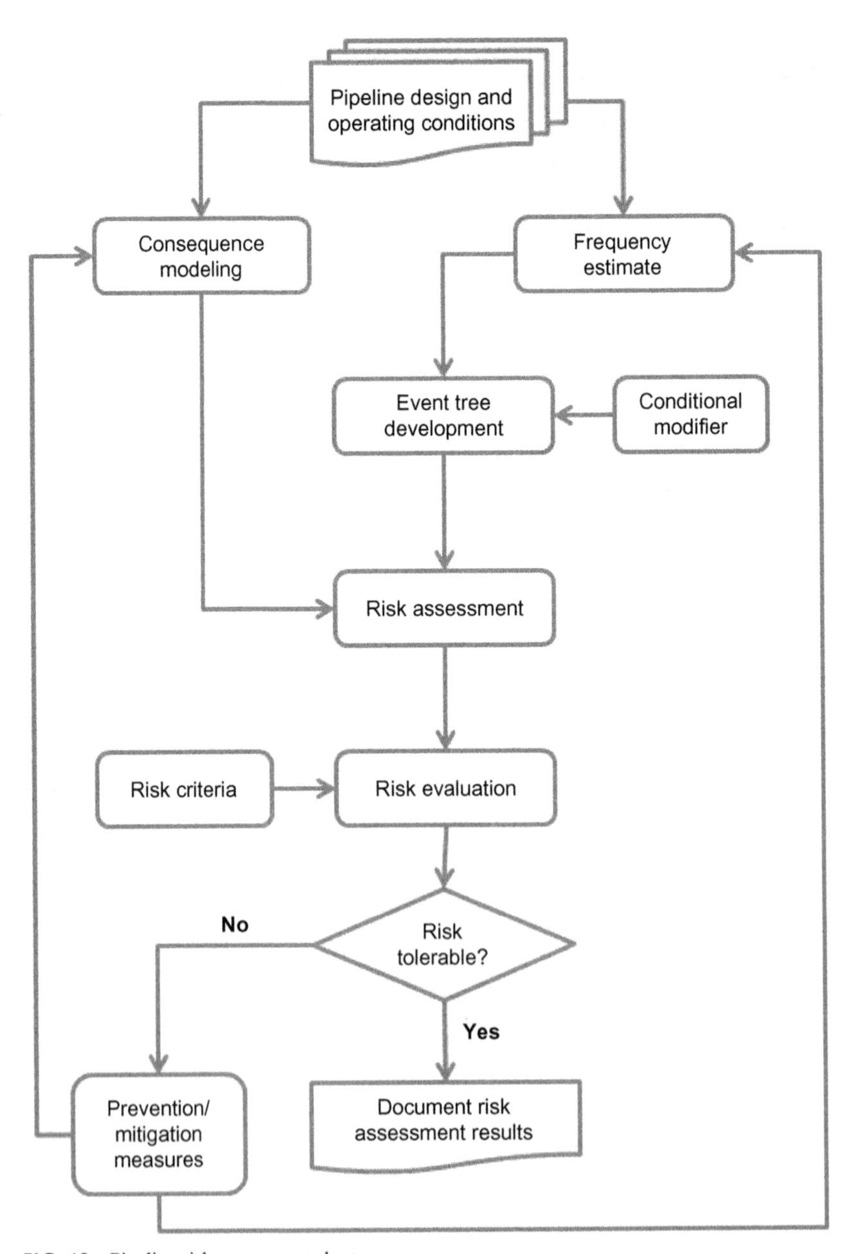

FIG. 12 Pipeline risk assessment chart.

conservative than the original model, and this could be due to the fact that not all information is available and therefore conservative assumptions were made as stated above. It also could be attributed in part to the fact that the current model was developed to be conservative as well. This case study proves that the current model is representative and can be used for pipeline risk assessments.

COMMERCIAL SOFTWARE

Commercial software is available in the market to conduct risk assessment, and they can be used for pipeline risk assessments. This book is not promoting any specific software, but a simple search can yield the available software and their features. The analyst could use these software if available, but the basics of risk assessment for pipelines were described in the book here. One should be aware of the limitations and assumptions implied in these software and packages when using them in order to make sure the results of the risk analysis are reflective of conditions of the pipeline being analyzed.

SUMMARY OF RISK ASSESSMENT

Risk assessment for pipelines is summarized in the chart shown in Fig. 12. It describes the different steps involved in the risk assessment as discussed in this chapter.

REFERENCES

[1] B. Rothwell, M. Stephens, in: Risk analysis of sweet natural gas pipelines: benchmarking simple consequence models, International Pipeline Conference, September 25–29, Calgary, Alberta, 2006.
[2] A. Shahriar, R. Sadiq, S. Tesfamariam, Risk analysis for oil & gas pipelines: a sustainability assessment approach using fuzzy based bow-tie analysis, J. Loss Prev. Process Ind. 25 (3) (2012) 505–523.
[3] Y.-D. Jo, B.J. Ahn, A method of quantitative risk assessment for transmission pipeline carrying natural gas, J. Hazard. Mater. 123 (1–3) (2005) 1–12.
[4] C. Vianello, G. Maschio, Quantitative risk assessment of the Italian gas distribution network, J. Loss Prev. Process Ind. 32 (2014) 5–17.
[5] S.B. da Cunha, A review of quantitative risk assessment of onshore pipelines, J. Loss Prev. Process Ind. 44 (2016) 282–298.
[6] J.C.A. Windhorst, Detailed Quantitative Risk Assessment of a Proposed Pipeline in Western Canada, IChemE.
[7] A.D. Little, Inc., in: An approach to the risk assessment of gasoline pipelines, Proceedings for the 1996 Pipeline Reliability Conference, Houston, TX, Gulf Publishing Co., Houston, TX, 1996.
[8] P. Hopkins, et al., in: Pipeline risk assessment: new guidelines, WTIA/APIA Welded Pipeline Symposium, Sydney, Australia, April, 2009.
[9] P. Tuft, N. Yoosef-Ghodsi, J. Bertram, in: Benchmarking pipeline risk assessment processes, International Pipeline Conference, Calgary, Alberta, September, 2012.
[10] UKOOA, Industry Guidelines on a Framework or Risk Related Decision Support, Issue No. 1, May, 1999.
[11] UKOOA, Guidelines for Quantitative Risk Assessment Uncertainty, Issue No. 1, March, 2000.
[12] Y.-D. Jo, K.-S. Park, B.J. Ahn, Risk Assessment For A High-Pressure Natural Gas Pipeline in an Urban Area, WIT Press, 2004.
[13] D. Kirchhoff, B. Doberstein, Pipeline risk assessment and risk acceptance criteria in the State of Sao Paulo, Brazil, Impact Assess. Project Appraisal 24 (3) (2006) 221–234.
[14] CCPS, Guidelines for Chemical Process Quantitative Risk Analysis, second ed., Wiley, New York, 2000.

[15] J.D. Rimington, in: Overview of risk assessment, International Conference on Risk Assessment, UK-HSE, London, October, 1992.
[16] UK Offshore Operators Association, UKOOA, Procedure for Formal Safety Assessment, Issue No. 1, November, 1990.
[17] Tripartite Symposium, Assessment of Major Hazards in the Process Industries—Review of Current Methods, University of South Wales, 1985. November.
[18] The Commission for Energy Regulation, ALARP Demonstration Guidance Document—Part of the Petroleum Safety Framework, Guideline No. CER/13/073, February, 2013.
[19] CCPS, Guidelines for Developing Quantitative Safety Risk Criteria, Wiley, 2009.
[20] UK-HSE, Societal Risk: Initial Briefing to Societal Risk Technical Advisory Group, Research Report RR703, 2009.
[21] N.J. Duijm, Acceptance criteria in Denmark and the EU, Danmarks Tekniske Universitet, 2009. Environmental Project No. 1269.
[22] U. Neunert, K.-D. Kaufmann, in: Presenting the societal risk of pipelines transporting hazardous materials, Symposium Series No. 158, IChemE Hazards XXIII, 2012.
[23] Hong Kong Risk Criteria, Available at: www.epd.gov.hk/eia/register/report/eiareport/eia_1252006/html/eiareport/Part3/Section13/Sec3_13.htm.
[24] CCPS, Layer of Protection Analysis: Simplified Process Risk Assessment, Wiley, 2010.
[25] Data collected from website: http://www.indexmundi.com/g/g.aspx?v=26&c=sf&c=tu&c=uk&c=us&c=xx&l=en.
[26] PHMSA Pipeline Risk Model working Group, Kiefner, in: General knowledge-paper study on risk tolerance, Presentation on October, 2016.
[27] HSE, Risk Criteria for Land Use Planning in the Vicinity of Major Industrial Hazards "Risk Criteria Document" (RCD), London, 1989.
[28] T. Maddison, The UK Approach to Land Use Planning in The Vicinity of Chemical Major Hazard Installations, UK-HSE.
[29] HSE Book, Risk Criteria for Land-Use Planning in the Vicinity of Major Industrial Hazards, 1989.
[30] Major Industrial Accidents Council of Canada, Risk-Based Land Use Planning Guidelines, Ottawa, first ed., 1995.
[31] J. Casal, C. Delvosalle, M. Demichela, Risk analysis, land use planning, Saf. Sci. 97 (2017) 1.
[32] US Transportation Research Board Special Report 281, Transmission Pipelines and Land Use, A Risk-Informed Approach, The National Academies Press, Washington, DC, 2004.
[33] This document contains Volume 2, Appendices of the 2007 Guidance for Protocol for School Site Risk Analysis. The entire guide is available at: http://www.cde.ca.gov/ls/fa/sf/protocol07.asp.
[34] Cornwell, J.B., and Marx, J.D., Application of Quantitative Risk Analysis to Code-Required Siting Studies Involving Hazardous Material Transportation Routes. Quest Consultants, Inc.

Chapter 7

Pipeline Process Safety Management

Pipelines are generally considered a safe mode to transport hazardous material in large quantities. Several assessments conducted to compare the risk associated with transporting comparable amount of hazardous material show that the risk is higher when using other methods such as railways or road trucks [1–3]. Generally speaking, loss of containment (LOC) incidents from rails are much more frequent. Some pipeline incidents can cause significant damage, but on average, rail incidents leak larger quantities per incident compared with the pipelines [3]. In conclusion, pipelines remain a safe mode of transporting hazardous material, but proper safety management programs must be implemented to maintain safety and reduce the risk associated with pipeline operations.

While pipelines remain safe, pipeline process incidents (LOC) cause damage, financial losses, injuries, and fatalities. Data collected for US network of pipelines are shown in Figs. 1–3. This information is based on the analysis of pipeline failure incidents of US DOT PHMSA. As such, there is still a strong need to implement proper process safety management (PSM) systems and programs to reduce the impact from pipeline operations.

PROCESS SAFETY VS. OCCUPATIONAL SAFETY

Occupational safety focuses mainly on preventing injuries to personnel that can result from incidents associated with the job they perform in the workplace. Requirements and regulations for this were established in the United States in 1970 through the Occupational Safety and Health Act. Occupational incidents tend to cause injuries to a small number of people. Common causes of nonfatal occupational incidents include the following [1]:

- Slipping, tripping, or falling incidents
- Being struck or compressed with an object or equipment (including motor vehicles)
- Motorized vehicle accidents
- Overexertion

Cross Country Pipeline Risk Assessments and Mitigation Strategies
https://doi.org/10.1016/B978-0-12-816007-7.00007-X

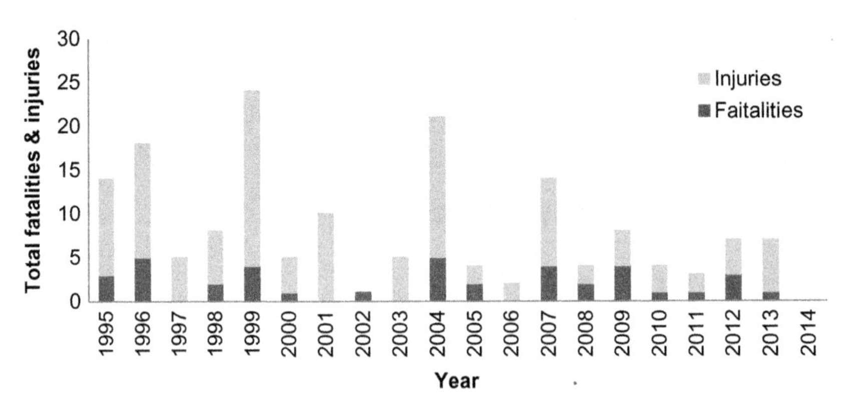

FIG. 1 US PHMSA data on hazardous liquid PL injuries and fatalities.

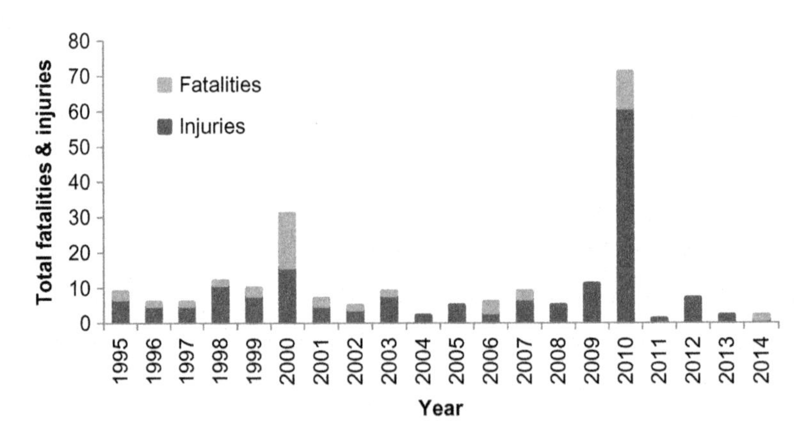

FIG. 2 US PHMSA data on gas transmission fatalities and injuries.

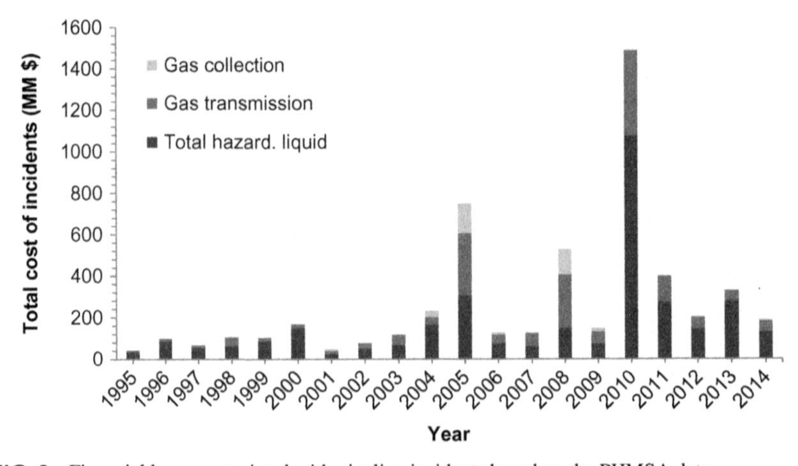

FIG. 3 Financial losses associated with pipeline incidents based on the PHMSA data.

Fatal injuries based on US DOL statistics show similar causes, as given below [2]:

- Violence and other injuries by persons or animals
- Transportation incidents
- Fires and explosions
- Falls, slips, and trips
- Exposure to harmful substances or environments
- Contact with objects and equipment

Process safety on the other hand focuses on preventing incidents related to the process that can mainly lead to loss-of-containment events. In other words, process safety for pipelines focuses on preventing leaks of hazardous material from pipelines that can cause catastrophic accidents and could potentially impact a large number of people. Process safety incidents typically cause a large number of fatalities and injuries compared with occupational safety events. A collection of the pipeline incident investigations, as well as recommendations and possible causes, for pipelines in North America are given at the website of the National Transportation Safety Board in the United States and the Transportation Safety Board of Canada. In the United States, PSM programs are defined by OSHA Standards 29 CFR 1910.119 (PSM of highly hazardous chemicals).

PIPELINE PROCESS SAFETY MANAGEMENT PRACTICES

Comprehensive description of PSM systems is available in the literature with great amount of details on the applications and frameworks available worldwide for PSM implementation [4–15]. In general, the main elements of PSM program (as applicable to pipelines) could be presented as shown in Fig. 4, based on the CCPS framework as presented in other references [16]. Implementing this program in pipeline operations should lead to safe operation and manageable risk. The same reference shows that other frameworks might have different elements or are organized differently but the objectives and the main elements are still the same.

The framework shown in Fig. 4 is based on four main focus areas, as explained below:

- Commitment process safety: this focus area including elements such as the following:
    - Compliance with standards and regulations that are often put in place to ensure proper PSM is being achieved through design, etc.
    - Reaching out to stakeholders (internal and external) such as employees and the public communities/partners to engage them in PSM and reflect their perspective on the issues of concern.
    - Ensuring and improving competency of all people involved in process safety through training and continuous development of expertise needed to ensure PSM is effective and working as expected.

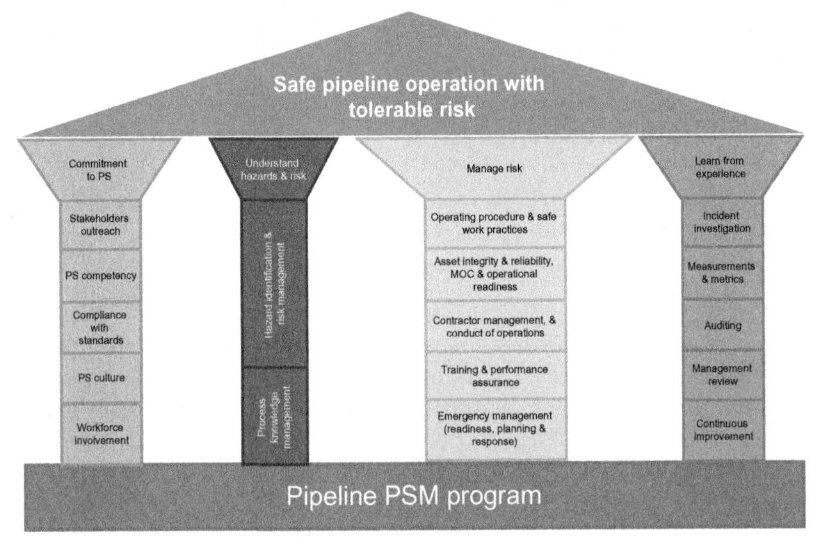

FIG. 4 Pipeline process safety management program based on CCPS process safety framework.

- Process-safety-oriented culture that promotes proper PSM practices and encourages active participation of all stakeholders in the process of managing process safety. Different studies are available on this issue that address the question on how to create and foster such culture at the workplace [4,5,7,9]. Employee empowerment and leadership commitment and accountability can enhance the proactive positive safety-oriented culture at the workplace [9].

- Understanding hazards and risk: this focus area requires process understanding in order to identify the potential hazards and conducting risk assessments. The scope of this book is related to this element and the next one.
- Managing the risk: this focus area include the following elements:
  - Adhering to operating procedure and implementing/complying with safe working practices including the effective implementation of work permit systems.
  - Maintaining the asset integrity through proper maintenance and inspection activities and programs and following proper/adequate Management of Change (MOC) processes to ensure all changes are properly assessed before introduced into the process.
  - Managing the safety of contractors working at the site and ensuring that they follow applicable safe working practices adopted by the organization.
  - Training the work force on the jobs expected to reduce the chances to introduce human error that can cause incidents. Human factors/errors are common contributing factors to process safety incidents.
  - Performance assurance through proper audits and implementation of performance indicators/metrics.

- Operational readiness and emergency management (i.e., planning, readiness, and response).
- Learning from experience: in this focus area, the pipeline or facility operator should seek continuous improvement through the following practices:
  - Investigating incidents and learning from incidents within the organization and from other applicable cases in the industry
  - Setting key performance indicators (KPIs) and other metrics to ensure the risk controls being implemented are working effectively
  - Conducting regular audit and compliance review to ensure all aspects of PSM programs are being complied with and adhered to (at least to the minimum level acceptable to ensure adequate/effective PSM at the organization)

The API RP-1173 presents a framework for pipeline PSM [4], which is consistent with the CCPS general framework referenced here, but it is presented in a different model that seems to focus on driving continuous improvement through the four main pillars of the PDCA model:

- Plan (P): this pillar focuses on risk management activities
- Do (D): this pillar includes the following elements:
  - Emergency preparedness
  - Operational control
  - Competence awareness and training
  - Documentation and record keeping
  - Stakeholder engagement
- Check (C): this pillar includes the following elements:
  - Incident investigation, evaluation, and lessons learned
  - Safety assurance
- Act (A): this pillar focuses mainly on the following:
  - Continuous improvement
  - Management review

At the heart of this model will lie leadership and management commitment. The PDCA is designed to encourage improvement and commitment to PS. Detailed description of the model is available in the API RP-1173 [4].

Both models shown here and many others rely on creating and fostering a safety-oriented culture, where safety is held as high priority among other factors. Without the proper safety culture in place, these models may not work effectively, and safety will not be prioritized. The set of attitudes, norms, actions, decisions, and messages delivered by leaderships of the organizations/companies should always emphasize the high-value safety holds in the organization. Three main elements will help create and sustain positive safety culture within the organization. These are the following:

- Stakeholders engagement
- Employee empowerment
- Leadership commitment to safety

PROCESS SAFETY INCIDENTS CAUSES

Generally, the main factors contributing the process safety incidents are attributed to the following points:

- Hazards not understood or not evaluated appropriately. This has been a major cause in several high-profile process safety incidents such as the explosion in Texas City Refinery in 2005 and the BLEVE incident of PETMEX in Mexico City in 1984 [17], as well as the Union Carbide plant incident in Bhopal, India (1984). In these cases, the hazard associated with the process was either not understood or not identified at all [18].
- Procedures are either not adequate or not being followed completely. This has contributed to some major incidents such as the Texas City Refinery explosion and Piper Alpha incident (1988). Bypassing procedures especially during operational transitions such as start-up or shutdown activities is common and can lead to major incidents. This is also applicable to deficiencies in work permit systems that also contributed to incidents as well.
- Inadequate emergency planning and preparedness. This was a major factor in PETMEX incident and the Union Carbide incidents as well. If emergency planning and preparedness is not adequate, process safety incident can escalate and increase the magnitude of damage caused by the initial incident [18].
- Improper design and faulty systems. This was a major contributing factor in Buncefield gasoline storage incident in the United Kingdom (2005) and in the Caribbean Petroleum Refining Tank Explosion and Fire (near San Juan, Puerto Rico in 2009). In both cases, it was found that the design of the level control/indicators on the storage tank was not adequate.
- Standards and regulations violation. Not adhering to applicable standards and regulation could cause process safety incident or escalate them leading to catastrophes. Violating standard requirements on spacing between hazardous facilities and the nearest public facilities contributed significantly to the disaster that follows the release of toxic gas in China in December of 2003 [17].
- Insufficient training and the lack of experience of the staff operating the facility do also contribute the process safety incidents. This was a contributing factor in the 2003 well blowout incident in China, for example, where the staff was not trained to recognize and handle the preincident events that could have been used to stop the escalation of events and prevent the catastrophe [18].
- Inadequate integrity management practices. If the mechanical integrity of the asset is not maintained at the required limits, it can fail and lead to serious process safety incidents. Corrosion management and inspection as well as preventative maintenance are essential to maintaining the integrity of the facility. Compromised mechanical integrity is among leading causes of process safety incidents as shown in surveys of major incidents in the industry.

- Management of Change (MOC). Hazards introduced during change to the process/facility design or operating conditions/procedures can be quite significant. As such, it is important to manage the change adequately to ensure no hazard is introduced without being assessed and mitigated. Any change shall be managed following well-established MOC processes and practices.

An adequate PSM program should take into account the lessons learned from previous incidents and have programs/processes in place to address these causes and reduce the likelihood of them contributing to incidents at the subject facility. There could be many other causes, but the ones mentioned above are quite common and seem to be repeated in many incidents. Therefore, it would be a good practice to address them even before incidents happen.

PROCESS SAFETY MANAGEMENT PERFORMANCE

In order to ensure that PSM program is working well, its performance must be monitored and tracked. The practical way to do so is by setting performance indicators and tracking them. Not all indicators are practical to track, so only KPIs would be needed. Otherwise, there will be a large number of indicators that are being monitored without adding a value. KPIs should be defined to measure the "health" or effectiveness of the PSM program and drive change to improve PSM.

The following parameters should be considered when setting process safety related KPIs:

- The KPI must be meaningful and measure something that can be measured.
- The KPI must be relevant to process safety and be linked to elements of process safety program (or active risk controls elements). KPI's definition should reflect what is intended to be measured.
- Data should be available and easy to collect. If it takes a huge amount of effort to collect that data, then it would be hard to implement the KPI especially if frequent update of the KPI is needed.
- KPIs have to be specific, in order to drive specific recommendations and lead to clear conclusions.
- KPIs should be designed to be updated with a sufficient frequency to ensure they provide the necessary level of performance monitoring and review.
- Ideally, KPIs shall be monitored and verified by an independent group integrity of the data being generated and reported.

KPIs can be either composite (composed of different parameters that can be lumped in one index) or can be based on a single parameter. There are also two types of KPIs, leading and lagging:

- Leading indicators: are indicators that are intended to provide an early warning about risk control or PSM element status and provide indication that it could lead to early intervention to improve PSM of the facility and

reduce the chance of failure before incidents happen. These are often referred to as Tier-3 and Tier-4 indicators per API RP-754 [19–21]. Tier-3 and Tier-4 indicators are indicators that are related to issues that represent "challenges to safety systems" or are related to "operating discipline and management systems" [19–21]. For example, relief valve failure would be Tier-3 event versus completion of inspections on time, which would be Tier-4 event [20].
Generally Tier-3 KPIs target would be zero, as it is intended to stop undesirable event. However, Tier-4 KPIs target would be 100% as it is intended for desirable task that would prevent incidents. "Tier-3 and Tier-4 KPIs are generally focused on monitoring and reviewing risk control systems" [21]. These could be engineering, administrative, or human-related controls and barriers. They also could be on both sides of the bow-tie diagram (i.e., prevention or mitigation measures).
- Lagging indicators: are indicators that indicate the failure of PSM systems in the form of incidents that happened. So, they tell about the incident that happened, which should provide "after the fact measure" of process safety status. They report LOC incidents in the unit of Process Safety Event (PSE) or PSE Rate (PSER). These are the number of incidents that happened to be normalized by the total man-hours of the working force (to be easy to compare it with other operators of different size), respectively. Tier-1 indicates much higher consequences that Tier-2 events. Release of large quantities of hazardous material would be Tier-1 event, while release of small quantities of hazardous material from its containment would probably be Tier-2. A good set of examples on how to define and assess pipeline PSM indicators and other cases are presented in the literature [5,22].

While Tier-1 and Tier-2 KPIs are clear and well defined and accepted in the industry, Tier-3 and Tier-4 require extra work to be defined in a manner that ensures their effectiveness. IOGP indicate that Tier-3 and Tier-4 KPIs should be

- proactive, based on hazard and risk analysis
- reactive, making use of lessons learned from incidents
- external learning from other similar organizations and applicable industry best practices

A step-by-step approach and guideline for developing process safety KPIs is presented in the literature [23]. A good number of leading KPIs are summarized in an industry survey that shows how popular specific KPIs are and at which management level are they being tracked in the organization [20,24].

API RP-1173 states that "pipeline operator shall establish and maintain a procedure to identify key performance indicators (KPIs) to measure the effectiveness of risk management and to improve pipeline safety performance" [4]. According to this recommended practice, KPIs should measure the health of pipeline PSM program:

- Lagging KPIs: this include a number of incidents and the resulting damage/losses from these incidents (LOC events)
- Leading KPIs should be established with proper reporting frequency and promote proactive approach to mitigating the risk. Process KPIs, such as the number of procedures modified/improved, are also to be defined.

SUMMARY

Pipelines well maintained and operated can be safer and have less environmental impacts than other modes of transportation. Proper PSM framework should be adopted for the pipeline based on international best practices to ensure that safety is maintained and risk is managed.

REFERENCES

[1] B. Westenhaus, Trucks, Trains, or Pipelines—The Best Way to Transport Petroleum, Oilprice.com, Aug 13, 2013.

[2] L. Young, Crude oil spills are bigger from trains than pipelines, Global News (2014). January 8.

[3] Fraser Institute, Safety in the Transportation of Oil and Gas, August, 2015.

[4] API, APR-RP 1173, Pipeline Safety Management Requirements, June, first ed., 2014.

[5] CCPS, Guidelines for Implementing Process Safety Management, second ed., Wiley, New York, NY, 2016.

[6] CCPS, Guideline for Risk Based Process Safety, Wiley, New York, 2007.

[7] C.A. Soczek, Implementation of Process Safety Management into Diverse Corporate Culture, DuPont Safety Resources, 2010.

[8] G.A. Papadakis, Assessment of requirements on safety management systems in EU regulations for the control of major hazard pipelines, J. Hazard. Mater. 78 (2000) 63–89.

[9] C.G. Blake, Kansas Corporation Commission Annual Seminar on API RP-1173, October, 2014.

[10] N.W. Hurst, et al., Measures of safety management performance and attitudes to safety at major hazard sites, J. Loss Prev. Process Ind. 9 (2) (1996) 161–172.

[11] G.B. DeWolf, Process safety management in the pipeline industry: parallels and differences between the pipeline integrity management (IMP) rule of the Office of Pipeline Safety and the PSM/RMP approach for process facilities, J. Hazard. Mater. 104 (2003) 169–192.

[12] B. Kichler, API RP 1173 Pipeline Safety Management Systems, PHMSA TQ.

[13] Energy Institute, High Level Framework for Process Safety Management, December, first ed., 2010.

[14] G.A. Papadakis, Assessment of requirements on safety management systems in EU regulations for the control of major hazard pipelines, J. Hazard. Mater. 78 (1–3) (2000) 63–89.

[15] G.B. DeWolf, Process safety management in the pipeline industry: parallels and differences between the pipeline integrity management (IMP) rule of the Office of Pipeline Safety and the PSM/RMP approach for process facilities, J. Hazard. Mater. 104 (1–3) (2003) 169–192.

[16] USA-DOL OSHA, Process Safety Management Booklet, 2000.

[17] W. Kent Muhlbauer, Pipeline Risk Management Manual: Ideas, Techniques, and Resources, Elsevier, third ed., 2004.

[18] A.M. Aloqaily, in: Industrial Disaster Management-Lesson Learned, Asia Process Safety Summit, 2014.

[19] J.A. Hare, M.P. Johnson, B. Fullam, in: Learning from process safety incidents, IChemE Symposium Series NO. 155, 2009, pp. 104–112. Hazards XXI.
[20] API Recommended Practice (RP) 754, Process Safety Performance Indicators for the Refining and Petrochemical Industries, Washington, DC, USA.
[21] IOGP, Report 556, Process Safety—Leading Key Performance Indicators, July, 2016.
[22] IOGP, Report 456, Process Safety Upstream PSE Examples, November 2011. (Updated November, 2016).
[23] Health and Safety Executive, Developing Process Safety Indicators, UK, 2006.
[24] CCPS, Process Safety Leading Indicators Industry Survey.

FURTHER READING

[25] Insurance Journal, Top 10 Causes of Workplace Injuries, January, 2015.
[26] USA-DOL, Census of Fatal Occupational Injuries (CFOI)—Current and Revised Data, 2015.
[27] Canadian Association of Petroleum Producers, Report on Process Safety Management: Regulatory Scan, Publication Number 2014-0026, 2014.
[28] CCPS, Process Safety Leading and Lagging Metrics, Revised January, 2011.

Index

Note: Page numbers followed by *f* indicate figures, and *t* indicate tables.

Q

R

S

T

U

V

W